THE LIMITS OF REDUCTIONISM IN BIOLOGY

The Novartis Foundation is an international scientific and educational charity (UK Registered Charity No. 313574). Known until September 1997 as the Ciba Foundation, it was established in 1947 by the CIBA company of Basle, which merged with Sandoz in 1996, to form Novartis. The Foundation operates independently in London under English trust law. It was formally opened on 22 June 1949.

The Foundation promotes the study and general knowledge of science and in particular encourages international co-operation in scientific research. To this end, it organizes internationally acclaimed meetings (typically eight symposia and allied open meetings, 15–20 discussion meetings, a public lecture and a public debate each year) and publishes eight books per year featuring the presented papers and discussions from the symposia. Although primarily an operational rather than a grant-making foundation, it awards bursaries to young scientists to attend the symposia and afterwards work for up to three months with one of the other participants.

The Foundation's headquarters at 41 Portland Place, London W1N 4BN, provide library facilities, open every weekday, to graduates in science and allied disciplines. The library is home to the Media Resource Service which offers journalists access to expertise on any scientific topic. Media relations are also strengthened by regular press conferences and book launches, and by articles prepared by the Foundation's Science Writer in Residence. The Foundation offers accommodation and meeting facilities to visiting scientists and their societies.

Information on all Foundation activities can be found at http://www.novartisfound.demon.co.uk

Novartis Foundation Symposium 213

THE LIMITS OF REDUCTIONISM IN BIOLOGY

1998

JOHN WILEY & SONS

Chichester · New York · Weinheim · Brisbane · Toronto · Singapore

Published in 1998 by John Wiley & Sons Ltd,
Baffins Lane, Chichester,
West Sussex PO19 1UD, England

National 01243 779777
International (+44) 1243 779777
e-mail (for orders and customer service enquiries): cs-books@wiley.co.uk
Visit our Home Page on http://www.wiley.co.uk
or http://www.wiley.com

Other Wiley Editorial Offices

John Wiley & Sons, Inc., 605 Third Avenue,
New York, NY 10158-0012, USA

WILEY-VCH Verlag GmbH, Pappelallee 3,
D-69469 Weinheim, Germany

Jacaranda Wiley Ltd, 33 Park Road, Milton,
Queensland 4064, Australia

John Wiley & Sons (Canada) Ltd, 22 Worcester Road,
Rexdale, Ontario M9W 1L1, Canada

John Wiley & Sons (Asia) Pte Ltd, 2 Clementi Loop #02-01,
Jin Xing Distripark, Singapore 129809

Novartis Foundation Symposium 213
ix+228 pages, 28 figures, 3 tables

Library of Congress Cataloging-in-Publication Data

The limits of reductionism in biology.
p. cm. – (Novartis Foundation symposium:213)
Papers from the Symposium on the Limits of Reductionism in Biology, held at the Novartis Foundation, London, May 13–15, 1997.
Editors, Gregory R. Bock (organizer) and Jamie A. Goode.
Includes bibliographical references and indexes.
ISBN 0-471-97770-5 (hbk:alk. paper)
1. Biology–Philosophy–Congresses. 2. Reductionism–Congresses.
I. Bock, Gregory. II. Goode, Jamie. III. Novartis Foundation for Gerontological Research. IV. Symposium on the Limits of Reductionism in Biology (1997 : Novartis Foundation) V. Series.
QH331.L554 1998
570′.1–dc21 98–2779
CIP

British Library Cataloguing in Publication Data

A catalogue record for this book is available from the British Library

ISBN 0 471 97770 5

Typeset in 10½ on 12½ pt Garamond by Dobbie Typesetting Limited, Tavistock, Devon.
Printed and bound in Great Britain by Biddles Ltd, Guildford and King's Lynn.
This book is printed on acid-free paper responsibly manufactured from sustainable forestry in which at least two trees are planted for each one used for paper production.

Contents

Symposium on The limits of reductionism in biology, held at the Novartis Foundation, London, 13–15 May 1997

Editors: Gregory R. Bock (Organizer) and Jamie A. Goode

This symposium was based on a proposal made by Lewis Wolpert and Ken Holmes

Participants

J. F. Ashmore Department of Physiology, University College London, Gower Street, London WC1E 6BT, UK

H. B. Barlow Physiological Laboratory, University of Cambridge, Cambridge CB2 3EG, UK

P. Bateson Sub-Department of Animal Behaviour, University of Cambridge, Madingley, Cambridge CB3 8AA, UK

D. Bray Department of Zoology, University of Cambridge, Downing Street, Cambridge CB2 3EJ, UK

S. Brenner Molecular Sciences Research Institute Inc., 9894 Genesee Avenue, La Jolla, CA 92037, USA

A. Burgen Department of Pharmacology, University of Cambridge, Tennis Court Road, Cambridge CB2 1QJ, UK

G. A. Dover Department of Genetics, 137 Adrian Building, University of Leicester, University Road, Leicester LE1 7RH, UK

D. Duboule Department of Zoology and Animal Biology, University of Geneva — Sciences III, 30 Quai Ernest-Ansermet, CH-1211 Geneva 4, Switzerland

A. Garcia-Bellido Department of Genetics, University of La Sapienza, Rome, Italy

J. Gray Department of Psychology, Institute of Psychiatry, De Crespigny Park, London SE5 8AF, UK

R. Henderson MRC Laboratory of Molecular Biology, Hills Road, Cambridge, CB2 2QH, UK

B. Hess Department of Biophysics, Max Planck Institute for Medical Research, Jahnstrasse 29, D-69120 Heidelberg, Germany

K. C. Holmes Department of Biophysics, Max Planck Institute for Medical Research, Jahnstrasse 29, 69120 Heidelberg, Germany

M. Kerszberg CNRS, Institut Pasteur, 25 Rue du Dr Roux, 75105 Paris, France

R. May Office of Science and Technology, Albany House, 94–98 Petty France, London SW1H 9ST, UK

J. Maynard Smith School of Biological Sciences, Biology Building, University of Sussex, Falmer, Brighton BN1 9QG, UK

G. Mitchison MRC Laboratory of Molecular Biology, Hills Road, Cambridge CB2 2QH, UK

M. Morgan Institute of Ophthalmology, University College London, 11–43 Bath Street, London EC1V 9EL, UK

T. Nagel New York University Law School, 40 Washington Square South, New York, NY 10012, USA

D. Noble Department of Physiology, University of Oxford, South Parks Road, Oxford OX1 3PT, UK

P. Nurse Imperial Cancer Research Fund, 44 Lincoln's Inn Fields, London WC2A 3PX, UK

M. F. Perutz MRC Laboratory of Molecular Biology, Hills Road, Cambridge CB2 2QH, UK

W. G. Quinn Department of Brain and Cognitive Sciences, Massachusetts Institute of Technology, Whitaker College, Building E25-436, Cambridge, MA 02139, USA

M. Raff MRC Molecular Cell Biology Laboratory, University College London, Gower Street, London WC1E 6BT, UK

S. Rose Department of Biology, The Open University, Milton Keynes MK7 6AA, UK

R. J. P. Williams Inorganic Chemistry Laboratory, University of Oxford, South Parks Road, Oxford OX1 3QR, UK

L. Wolpert (*Chairman*) Department of Anatomy and Developmental Biology, University College London Medical School, Medawar Building, Gower Street, London WC1E 6BT, UK

Introduction

Lewis Wolpert

Department of Anatomy and Developmental Biology, University College London Medical School, Medawar Building, Gower Street, London WC1E 6BT, UK

> 'If one had a proper knowledge of all the parts of the semen of some species of animal in particular, for example of man, one might be able to deduce the whole form and configuration of each of its members from this alone, by means of entirely mathematical and certain arguments.'
> R. *Descartes — On the Formation of the Fetus*

Perhaps I should begin by explaining the origins of this symposium. I am not a philosopher, but I began to think about reductionism when I attended a meeting at which Ian Stewart, the mathematician, talked about something called the 'Taylor cell' in hydrodynamics. In this system there are two glass cylinders, with one inside the other such that there is a thin layer of water between the two. The inner cylinder is rotated, and at a certain speed under certain conditions the water breaks up into a series of dark stripes. If you write down the Navier–Stokes equations for fluid motion and put in the boundary conditions, and enter these into the computer, this behaviour can be fully predicted. Does this mean to say that we have understood it? His argument was that in order to make sense of this system one has to do a meta analysis of the equations. I was left puzzled: does the fact that one can write down the equation mean to say we have understood the process?

Then I went to Ken Holmes' birthday party, which is really the origin of this meeting. When Ken told me how much he knew about muscle, the thought occurred to me: how much more does he want to know? Does he want to know about muscle action at the level of quantum mechanics? To some extent he seems already to have solved the problem. I have also spoken to crystallographers who have told me that we will never be able to predict the folding of proteins from first principles.

Then, in a different context, I interviewed Roald Hoffman, the Nobel laureate in chemistry. To my amazement, here was a man who has claimed to apply quantum mechanics to chemistry yet he is militantly antireductionist. His idea is that as we reduce chemistry to physics, we lose the chemistry.

Along other lines, I'm well known for my depression, and consequently this is a subject area I have great interest in. This has raised for me deep questions about what it means to try to understand depression. I don't even know what language to use or at what level of organization to look to try to answer this problem.

As a final example of the sort of problem we are facing in attempting to trace the limits of reductionism, I recently went to a lecture by Tony Hunter given at the Royal Society. He stated that 5% of the genes in our body encode tyrosine kinase receptors—that's an awful lot. Then he went on to say that half the genes that we have are involved in intracellular communication: that is, membrane transduction. I began to think: how are we ever going to understand what's going on? And do we really want to know all that is going on? In other words, to what level do we want to reduce all this complexity?

Of course, the idea of levels of organization is an old one and I'm sure it is going to dominate a great deal of this meeting. I came across a nice article from 1972 by Philip Anderson, the physicist, who said that if one really is a reductionist and believes that there's a theory of everything, we should just work on that—the rest is trivial. If everything can be explained in terms of fundamental physics, why work on anything else? He claimed that this position is obviously ridiculous. He also pointed out that fundamental physics has contributed nothing to biology. My physicist friend Michael Fisher maintains that one way of looking at a field in terms of reductionism and levels of organization is to ask: am I at all interested in anything going on in the level below? In other words, am I as a developmental biologist concerned with what's going on in cell biology? Unquestionably, yes. In the molecular biology? Absolutely. Am I interested in advances in chemistry? Not one hoot. Quantum mechanics or solid-state physics are irrelevant to developmental biology. Thus there is this problem about how one relates the different levels. It seems to me that a key feature of moving from one level to another, and why it does matter, is that we are not allowed to have anything at one level that is inconsistent with something at another. For instance, we are not allowed to explain anything at the level of cell biology which is forbidden by chemistry. Otherwise you could have a level of organization dealing exclusively with the paranormal that is beautifully internally consistent, but if you can't relate it to any other level then I would say it is not science.

I have been amazed how much interest this subject generates and how many people have strongly held views on reductionism. One of the things that people care about is the higher levels of biology: processes such as emotion and consciousness. I hope we won't spend too much time discussing consciousness: it's a word that I believe should only be used with a permit and then only rarely. If we all reach a consensus I'll be amazed, but it would be nice if during the course of the next few days we could work out what the real problems are, and perhaps even change our minds.

Reductionism and antireductionism

Thomas Nagel

New York University Law School, 40 Washington Square South, New York, NY 10012, USA

Abstract. Reductionism is the idea that all of the complex and apparently disparate things we observe in the world can be explained in terms of universal principles governing their common ultimate constituents: that physics is the theory of everything. Antireductionism comes in two varieties: epistemological and ontological. Epistemological antireductionism holds that, given our finite mental capacities, we would not be able to grasp the ultimate physical explanation of many complex phenomena even if we knew the laws governing their ultimate constituents. Therefore we will always need special sciences like biology, which use more manageable descriptions. There may be controversy about which special sciences cannot be replaced by reduction, but that there will be some is uncontroversial. Ontological antireductionism holds, much more controversially, that certain higher-order phenomena cannot even in principle be fully explained by physics, but require additional principles that are not entailed by the laws governing the basic constituents. With respect to biology, the question is whether the existence and operation of highly complex functionally organized systems, and the appearance of self-replicating systems in the universe, can be accounted for in terms of particle physics alone, or whether they require independent principles of order.

1998 The limits of reductionism in biology. Wiley, Chichester (Novartis Foundation Symposium 213) p 3–14

The reductionist idea, which was already a gleam in the eye of the Presocratic philosophers, and which has been such a creative driving force in the history of modern science, is this: All of the complex and varied and apparently disparate things and processes that we observe in the world can be explained in terms of universal principles that govern the common ultimate constituents out of which, in many different combinations, those diverse phenomena are really composed. The idea is that there exists, in principle, a theory of everything, in the form of a theory governing the one thing or few things of which everything else consists. Let me refer to this as the *ultimate* level of explanation. Reductionism has two aspects: constitutive and explanatory. The constitutive thesis is that everything is made of the same elements; the explanatory thesis is that everything that happens can be given an ultimate explanation in terms of the laws governing those elements.

There are two kinds of *anti*reductionism: epistemological and ontological— having to do, respectively, with what we can know and what is really the case.

Epistemological antireductionism holds that even if *in reality* everything is explained by particle physics, we cannot, given our finite mental capacities, grasp the ultimate explanation of most complex phenomena, and would not be able to do so even if we knew the law or laws governing their ultimate constituents. We are therefore constrained to make do with rougher explanations couched in terms that our minds can accommodate.

With regard to many of the things that go on in the world, epistemological antireductionism is uncontroversial. If there were a detailed causal explanation at the subatomic level of the movement of a pianist's fingers in the course of playing a Beethoven sonata, we would not be able to hold it in our minds. The same is true of the ultimate explanation of a thunderstorm, or of a rise in the kangaroo population of Australia. Analysed into their ultimate constituents, these phenomena are too complicated for us to handle. Moreover, ultimate explanation of the precise physical event in each case would not tell us what we want to know. The understanding we want requires explanations that refer to features of those events that we can observe or grasp — explanations in terms of music, or meteorology, or population ecology. The idea of the ultimate explanation of such things must remain, at best, a pure idea, and the same can probably be said of most familiar phenomena.

Still, the pursuit of reduction is often successful and it goes in stages. Even if there are some things that we can be sure will never be epistemologically reducible to the ultimate level, there are levels in between, such as those of traditional chemistry or atomic physics, and development can occur over time in our capacity to explain complex phenomena in more basic terms. The 19th century chemistry summarized in the periodic table of the elements was explained eventually in terms of a physics of atoms composed of protons, neutrons and electrons, and the descent to more basic levels has continued, so that reductions are epistemologically possible now which were not in the past. Even so, explanations at higher levels often remain practically preferable and for many purposes indispensable.

Others directly involved in these sciences will be discussing the epistemological problem, but I want to concentrate on ontological antireductionism, since it presents special philosophical problems. It can take more than one form, but it is always a claim about what the world is like, not just about certain conditions or limits on our knowledge of the world. What does the natural world, including ourselves, *really* consist of — as opposed to how we with our limited minds and practical needs find it convenient to describe it?

It is easiest to explore reductionism by considering its denial, and ontological antireductionism can take two forms — constitutive and explanatory — corresponding to the two aspects of antireductionism already mentioned. I won't say much about constitutive antireductionism. Its most famous example is the psychophysical dualism of Descartes—the position that not everything in the world is constituted of the same basic elements, those studied by physics, but that there are non-physical events or things as well—conscious mental events and conscious subjects. That remains an interesting issue, but it is distinct from the broader question of the limits of reductionism in biology, which is our topic, so I shall leave it aside. Nor shall I discuss the form of vitalism that holds that all living organisms, conscious or not, contain a non-material vital principle in addition to the universal chemical elements. I suppose no one believes that today.

My topic will be the ontological issue between reductionism and antireductionism at the level of explanation. The ontological antireductionism I want to discuss has to do with the physical world, including the biological organisms in it, conceived as complex physical systems. It is the position that some physical phenomena, even though they can be explained in terms of principles that fit their specific features, simply do not have an explanation at the ultimate level—that is, in terms of the universal laws governing their ultimate constituents.

For the purpose of examining this type of view, I shall simply assume that everything from stars to organisms is composed of the same ultimate common constituents and that those constituents behave in conformity with universal laws. It is clear that for practical purposes, much of our understanding of living things must stay close to the biological level. Even though new levels of explanation become available over time, they do not necessarily result in the elimination of the old. For example, I gather that explanations of heredity in terms of classical genetics, descended from the Mendelian theory, are not about to be simply replaced by explanations in the language of molecular biology (Kitcher 1984). But whatever the epistemological situation, there remains a question about the ultimate facts: Is the fullest explanation of all physical phenomena, including those of biology, the one that refers only to the most basic and universal physical level?

The question is sometimes put by asking whether there are *emergent* laws of nature—laws that emerge only at a certain level of complexity and that govern the behaviour of complex systems without being derivable from the laws that govern their simple constituents. One of the problems with this view is that whenever we discover apparently emergent laws, provided they are sufficiently precise, we have to consider whether they are emergent only relative to our present physical theory, or more absolutely. Resting content with emergence is contrary to the scientific impulse: we tend to try to postulate previously unknown properties of the constituents which account for the laws reductively after all. This may or may not turn out to be possible, but the tendency shows that the reductionist aim operates as a norm setting the ideal form of understanding toward which we hope we are heading.

Nevertheless, since emergence is the main alternative to reduction, it merits exploration. Emergence might rely on a supposition of indeterminism at the level of particle physics. According to this possibility, indeterminism in basic physics leaves some things unexplained which are nevertheless explicable by principles that govern phenomena at higher levels of complexity, and perhaps govern the development of complexity as well: so while the existence and behaviour of such higher-order systems is consistent with the basic laws of physics that govern their ultimate constituents, they are not causally determined by those laws to the degree to which they are causally determined by higher-order laws. Relative to basic physics, in other words, such systems are purely possible but vanishingly unlikely, yet they are predictable by reference to higher-order principles. I won't say more about this, both because the supposition of indeterminacy is problematic (it arises only at the level of measurement, not of the basic equations) and because I believe the problem of emergence can be posed even about a physically deterministic world.

The question is, what would it mean for a higher-level phenomenon to be explicable *only* in higher-level terms? Suppose that there is, whether we can know it or not, an ultimate explanation of every particular thing that happens in the physical world — an explanation in terms of the laws governing its basic physical constituents that makes it as determinate as it can be made by any higher-level explanation. Suppose that this is true of everything from thunderstorms to the performance of piano sonatas — or at least the actual physical events by which a particular performance came about. Still, it might be the case that this ultimate explanation of the individual event in all its detail left something unexplained: namely why *that type of higher-level event* occurred — why the pianist hit those notes rather than why that specific chemical event occurred. To understand why he hit the notes we need a cruder level of description of the antecedent circumstances that would produce that type of result, even if the physical details had been different, in both the causes and the effects.

Of course if there is an ultimate explanation of a specific physical event, and that event *is* an event of a higher-level type, then in a sense the ultimate explanation explains why an event of that type occurred. But it does not explain why *some event or other* of that type occurred, and that is often what we want to know. Explanation has to seek its own level (Garfinkel 1981). Events of certain physically non-basic types, such as a tree shedding its leaves or a person paying a phone bill, can occur in many different ways — that is, they can be instantiated by many different complex events that are distinguishable at the basic physical level. To explain why *some* event of that type occurred, one must find conditions and principles that fit the type, and not just its specific instantiation on that occasion.

The fact that we are interested in the explanation of a higher-level event does not guarantee that there is one: If a father wants to know why all three of his sons were killed in the war, there may be no answer except a separate explanation of how each of them was killed. But sometimes we do find higher-order laws governing complex events — in economics, population ecology, the psychology of conditioning, or (I am told) thermodynamics — without being able to derive them from particle physics. By itself, this doesn't show whether the higher-order laws are irreducibly emergent, or only emergent relative to our present knowledge. We have reason to *look* for reductive explanations, level by level, for such apparently emergent principles, since it is contrary to the spirit of scientific inquiry to leave black boxes permanently unopened.

Still, the possibility of explanatory irreducibility is relevant to the question of the limits of reduction in biology. If there really are biological explanations that accomplish something that cannot even in principle be done at more basic levels, chemical or physical, that would be an important limit to reductionism, not just a limit on our knowledge but a fact about the world. The issue has been discussed by philosophers under the heading of the status of the so-called 'special sciences' (Fodor 1974). Do the truths of all the special sciences — sciences that deal only with a restricted subset of natural phenomena, such as biology, psychology, or economics (if it is a science) — derive from physics? This question is particularly acute with respect to the life sciences; there seems less reason to doubt the derivability in principle of geology or astronomy from physics.

How could the following two propositions both be true?

(1) Every event that happens in the world has a fundamental physical description and a fundamental physical explanation.
(2) Some facts about the world do not have a fundamental physical explanation, but do have a higher-level explanation.

The answer is that they could both be true if the higher-level explanations depended on principles governing the relations between general types of phenomena or properties that were not subject to correspondingly general characterization in ultimate physical terms, even though each instance of such a phenomenon had a distinct ultimate physical characterization. Perhaps not all naturally important kinds correspond to kinds definable in basic physics. If that were so, the laws operating at the higher level could not be derived from corresponding laws couched at the fundamental level, even though each event falling under the higher-level laws could be given a separate ultimate explanation.

Every science seeks interesting general principles that support counterfactuals (i.e. statements about what would have happened if things had been different), not just detailed descriptions of individual events. Science tries to discover how things work in general, so that if we know why something happened, that tells us that something like it would have happened under other, relevantly similar conditions, but that if conditions had been relevantly dissimilar, it would not have happened. So the dimensions of relevant similarity are crucial in the formulation of scientific hypotheses. Are there, then, some sciences for which the relevant kinds or similarity dimensions are irreducibly higher-level and not equivalent to similarities in the terms of a more basic science — or the most basic and universal of all?

The clearest examples of important properties that have no basic physical equivalent are not natural but conventional. The contingency of conventions of language and money make it self-evident that there is no purely physical equivalent of someone's asserting that the Earth is round, or of someone's rent having gone up 20% in the past year. Explanation in psychology and the social sciences is completely dependent on generalizations that refer to such physically irreducible properties. Since conventions depend on the contingent practices of human societies, however, that doesn't tell us much about the issue of reduction in the natural sciences. If there are important natural kinds that are not equivalent to basic physical kinds, it will not be because human conventions determine the ways in which they are physically instantiated. Conventions do not carve nature at the joints.

One concept specifically relevant to biology, however, is that of a *functional* property, and functional properties too have no basic physical equivalent. For a very simple but still non-natural illustration, consider the property of being in 'shift' position, as this applies to any keyboard writing device. It is a functional property because it is defined in terms of its role in relation to the conditional inputs and outputs of the device, rather than in terms of its intrinsic physical character, and it explains why when you strike a key, you produce a capital letter. In an old-fashioned manual typewriter, it is physically realized by the raising of the carriage by a key that acts as a lever, so that the upper half of each key hits the ribbon. In an IBM Selectric, it is physically realized by a 180° rotation of the ball. In my word processor, God knows how it is physically realized. Now the functional properties of an artefact are not candidates for ontological irreducibility, because even though they are not equivalent to natural physical kinds, neither are they natural kinds of any other sort: *they don't really explain how the world works*, because artefacts are constructed by human beings to work as they do on the basis of the physical properties of their constituents.

The interesting question is whether there are functional properties *in nature* that do not correspond to natural kinds of basic physics, but that really do enter into generalizations that physics can't explain. There certainly are functional properties in nature, such as that of being a wing or an eye or a gene or a muscle, and epistemologically they are essential to our understanding. But at this point we run into a reductionist response parallel to the point just made about artefacts: namely that the functional properties of biological systems are the products of evolution by natural selection among alternatives whose availability is antecedently given by the physical properties of their constituents. That a neuron or an eyeball or a gene will work the way it does is therefore just as strictly derivable from pure physics as is the operation of a typewriter. In both cases, the functional properties may be indispensable to our practical understanding, but the real causality is physical, not functional. The theory of natural selection means that all reference to purpose or function in nature is misleading, because it has to be taken as shorthand for an entirely mechanistic historical explanation.

That is the conventional wisdom, at least among the vulgar. But is it true? It follows from the application of an undeniably attractive reductionist world picture—the idea that evolution by natural selection completely bridges the gap between particle physics and biology (Kauffman 1993).

This makes evolutionary biology radically reductive, and much harder to really absorb than is generally realized, because we are allowed to keep our old purposive explanations, with the vague qualification that they all have to be reinterpreted in terms of survival value. But are there more specific reasons to be convinced that the range of biological possibilities available for natural selection is simply a logical consequence of the most basic principles of physics?

There are two separate questions here. First, do the basic laws of physics explain the functional organization of organisms, i.e. the principles which specify in higher-level biological terms what a given type of system will do under different conditions, internal and external? Second, is the coming into existence of such organisms either entailed or rendered probable or anyway not vanishingly improbable by basic physics, given the existence of the physical cosmos? The existence of typewriters is something whose *possibility* is entailed by basic physics, but typewriters exist only because people have made them. If the answer to the first question is yes, so that the *possibility* of the existence of eyes and ovaries is entailed by physics, then the full ontological reducibility of biology to physics depends on the reducibility of the evolutionary process, without recourse to additional basic principles of higher-level organization. And it depends not just on there being a full physical account of the exact course of events which has actually occurred, but an account of why some such development of complex systems was likely or not unlikely to occur independent of the precise initial cosmic conditions, physically specified. This would have to include, of course, an account of why physics makes likely the coming into existence of the most important functional property of all, the property possessed by certain complex systems of being *self-replicating*.

To sum up: Epistemologically and practically, there are obvious limits to the level of reduction we can hope to achieve in biology, or that would be useful. Higher-order understanding will remain indispensable. As for the ontological question—what the world is really like—we just don't know how far reduction may go, but we can be sure that the search for ever more reductive explanations will continue to be one of the most powerful motives of science.

References

Fodor J 1974 Special sciences, or the disunity of science as a working hypothesis. Synthese 28:77–115
Garfinkel A 1981 Forms of explanation. Yale University Press, London
Kauffman S 1993 The origins of order. Oxford University Press, New York
Kitcher P 1984 1953 and all that: a tale of two sciences. Philos Rev 93:335–373

DISCUSSION

Rose: I wanted to query some unexplained terms. Both you and Lewis Wolpert (in his introduction) used the term 'level'. What constitutes a level? Do you see this as an ontological or epistemological description of an aspect of the natural world? Secondly, you made a distinction between the terms 'explanation' and 'description'. I don't understand why a basic physical account of a phenomenon is an explanation, whereas a functional account, for instance of the role of a wing, or of money, is only a description. And the third issue that I don't understand is why one should automatically assume that if you can describe a phenomenon in physical terms this is in some way more 'fundamental' than describing it in terms of functions of a system.

Nagel: It is an open question whether levels are epistemological or ontological. Take the familiar distinction between chemistry and physics: this is an example of a successful reduction. In advance of the explanation of chemistry by physics it was possible to distinguish between those two levels and perhaps not to know the extent to which the first could be explained in terms of the second. Thus it is possible that a distinction between levels can be epistemological, and it is possible that it can be ontological: only the success of a reductive explanation will answer the question.

With regard to 'explanation', I simply use the word 'fundamental' to express the idea behind reductionism: an explanation of an event in terms of the principles governing its most basic constituents—those that it shares with everything else in the universe—can be called 'fundamental'. It may well be that an explanation of this kind, even if it exists, is inaccessible to us. Even if it is accessible, it may be useless for the purposes we need it for. An explanation has to be tailored to what we want to know about the event. Taking up the analogy of the piano player, if we want to know why he played those notes rather than why the muscles of his hands moved in a particular way, then we will want antecedent conditions which, if they had been different at that level, would have led to a different outcome at the level of description at which the explanation is sought. Explanation is correlative with the kind of description of the thing to be explained.

Bray: In simple terms, understanding is a human attribute, not something that is intrinsic to the system. When I understand something, it is because I have a small number of mental objects which I'm able to manipulate internally according to well defined rules. This process provides a simple explanation for what I observe: in that sense it is a language.

I would like to raise the question of whether our intellects are expanding into the realm of computers. Is this going to enlarge our capacity for understanding, and thus change the levels at which we view reductionism?

Noble: The sense I get, whether correctly or not, from what Thomas Nagel has said, is to wonder seriously about the initial distinction between the epistemological and the ontological. If we identify a type and it is such that we are forced to say that because of the nature of that type we can't refer to there being fundamental, lower-level explanations of what is going on because that type can't be analysed or referred to at that level, aren't we saying also that there is something there at the higher level? That is to move over into ontology: it actually exists.

Nagel: I used the analogy of the typewriter. We can say it printed a capital letter because the shift key was depressed, but that's not really how the universe works. Typewriters in shift position are not real aspects of the causal order. Not everything that we can say about the world has the kind of reality that science is after. There is a real question in biology about the explanatory status of the functional level of explanation, which is comparable. Of course wings exist and eyes exist, but I was just trying to explain the opposition between the point of view which says that while we have to see the world in this way it really is all physics, and the view that what we see here is a part of the way nature is carved up.

Gray: I want to go back to something that Lewis Wolpert said in his introduction, which I think is an important distinction to keep in mind, namely the contrast between that which the laws at a lower level permit at a higher level, and that which the laws at a lower level impose or require at a higher level. This tension between physics and biology, or between the physical law explanation and the functional explanation, centres exactly around that distinction. If you consider the way natural selection works, it can only work if you have variation in a manner which we can describe quite effectively in terms of information. The analogies about libraries and the information contained in DNA sequences are familiar to all of us. You can only have an information system if you have degrees of freedom. If everything were determined fully by laws other than the selection process that is acting upon that information system, you couldn't have the variability in which the information consists. Coming down to the physicochemical basics, in so far as I understand them, you can have selection operating upon the bases that make up DNA precisely because the physicochemical constraints permit several possibilities at each point. Natural selection can then work on those possibilities without violating the laws of chemistry and physics, but by adding something to those laws which is not itself explicable in terms of those laws alone, but is explicable in terms of the interaction between organisms and their environment. You can of course then say, when you look at any specific organism and its interaction with the environment, that everything is going on in a way which requires, and for which there can be given, a detailed physical explanation of each event in that interaction. But, none the less, in order to understand how functions work, it is necessary to bring in these further understandings from an information perspective, and this is not itself a physicochemical perspective.

Quinn: As a practising scientist it has been my experience that sometimes things work and other times they don't. But reductionism is always the thing you try first—it is like Pascal's wager: if you believe that the universe isn't explicable, you're dead before you start. If you believe that it is explicable and you try and explain it you might fail, but at least it's a try. A well structured testable wrong idea is better than a vague vaguely right idea. The extreme opposite of reductionism is complete description: Linnaeus versus Darwin.

Wolpert: Are you really saying, no reductionism, no science?

Quinn: No. What I'm saying is that anybody coming into a new field tries reductionism first: sometimes it doesn't work and you then try an alternative approach.

Holmes: Concerning the problem of explanation, in physics and biology there are apparently rather different expectations as to what constitutes an explanation. In principle, at least in physics, everything should be reducible to equations of motion or field equations, which are predictive. In biology we only have one theory, a historical theory which is in no grand sense predictive. Having said this, it is clear that global prediction of complex physical systems is in fact illusory so that physics and biology are actually reduced to doing the same thing: namely, looking at subsystems and doing essentially what every engineer does, solving small systems pragmatically using free parameters. Ultimately, therefore, explanation is something we use in a local sense quite effectively, but in these global senses actually lacks meaning.

Wolpert: Can you give an example of this?

Holmes: Let's say we can use the Schrödinger equation to get as far as a hydrogen atom, but as a many bodied problem we already go wrong, so we give up. In biology, we know only of Darwin's theory and we cannot predict with this. Again, we reduce to doing something else: to explain by defining subsystems which you can analyse with a small number of parameters, each chosen for their usability or measurability.

Gray: There was a recent paper about the changing lengths of lizards' legs which was precisely a predictive application of Darwinian theory (Losos et al 1997).

Williams: My point is very simple really. If we take our thinking back to the beginning of the Universe and start from the big bang, we have a problem with the development of it shortly afterwards in that it did not expand just radially, but it gained angular momentum, which is the introduction of turbulence. Where does turbulence come from? Your reductionist principles have gone, because you have no idea where turbulence comes from. If you start from a point in the big bang, everything goes outwards on a radius; nothing goes round. In turbulence, things are going round but where did such motion come from? The point follows that if there is a one-off event, then we have a hard job conducting a scientific investigation or discussion of it. This means we can't reduce past the one-off events: the big bang, the introduction of turbulence, the origin of life.

Nurse: This issue of explanation versus description is interesting. I think of it differently to Thomas Nagel. For me there is often a greater sense of explanation when working at a higher level and more a sense of description when working at a lower level. Discussing a wing in terms of its function provides more of an explanation of a 'wing' than describing its molecular structure. This may reflect how humans think: that is, we only find satisfactory explanations when thinking about a phenomenon at levels close to the level at which the phenomenon operates.

My second point has to do with this issue of new laws coming into place at new levels. You mentioned the concepts of souls and of vitalism. There can be a lot of baggage that comes along with concepts such as these: when the word 'soul' is mentioned one immediately starts thinking of things that are not explicable by the laws of physics.

Morgan: My comments follow on because they are concerned with the cognitive issue. I wonder whether we should try to dissect this epistemological issue further. The proposal is that there are limits to reductionism that are imposed by the nature of our cognition and our minds. However, these could be of two very different varieties. One could be just to do with the amount of information that we can process: taking your example of a thunderstorm, suppose in principle that we could describe a thunderstorm by the movement of every particle within it. Clearly one reason why that would be useless is that there would be too much information for us to take on board. On the other hand, suppose one could derive a satisfactory equation describing those particles, that might still defeat our imagination in the way that natural selection defeated the imagination of many of Darwin's contemporaries, and the way in which the general theory of relativity still defeats the imagination of most people. These are entirely different forms of cognitive limit. Which of the two did you have in mind when you proposed epistemological antireductionism?

Nagel: The epistemological/ontological distinction has been at the centre of much of this discussion. It is true that we don't want an understanding of everything in terms of particle physics. We wouldn't feel we understood biological processes if someone were to give us a particle physics account of them because we don't see the world in those terms: instead, we want to understand the world in the categories in which we perceive it and carve it up naturally. But I still think there remains the question of whether the real truth about causal order of the universe is somewhat different from what is easiest for us to grasp. Thus there might a reductionism which simply doesn't conform to the shape of the science that makes things comprehensible to us.

Reference

Losos JB, Warheit KI, Schoener TW 1997 Adaptive differentiation following experimental island colonization in *Aniolis* lizards. Nature 387:70–73

Reductionism in physical sciences

R. J. P. Williams

Inorganic Chemistry Laboratory, University of Oxford, South Parks Road, Oxford OX1 3QR, UK

Abstract. The idea of reductionism in physical sciences is that all physicochemical observables can be described in terms of a limited number of particles and their variable energies. Here we limit ourselves to atomic descriptions showing how very successful reductionism is in treating equilibrium systems. This includes all properties of single molecules, even DNA, and can be extended to dynamic assemblies of molecules through the variables composition, potential energies, kinetic energies (temperature) and volume (pressure). This description includes the capacity of a system to change, to do work. It does not include working or changing systems when we have to consider time-dependent variables such as directed motion, flow. Analysis of such accidentally or purposefully directed activity seems, to the author, to be outside the above reductionist analysis in that its feature is organization around a 'plan' or a 'cycle'. Thus reductionism fails to describe machines, man-made or biological, in that the parts are arranged, even dedicated, to a total function.

1998 The limits of reductionism in biology. Wiley, Chichester (Novartis Foundation Symposium 213) p 15–35

The central tenet of reductionism is that a complex system can be *fully reduced* in terms of its small component parts. It is implied that given the correct circumstances the component units will reform the whole system without additional input or aid. As it stands this leads some authors (e.g. Atkins 1995) to attempt an infinite reductive analysis of the whole universe into a theory of everything based on the mathematical analysis of 'particles' and 'energy'. I am sure that nobody in this book wishes to discuss such an approach. In any event even in mathematics it is clear that certain shapes cannot be reduced to smaller parts, e.g. a Möbius ring or a knot (Penrose 1995). My analysis of systems must be limited to their easily appreciated small components of well-defined character and their associated variables. I shall try to show the successes and failures of reductionism.

The classical reductionism of the Greeks and Chinese classified all material around us into four component 'elements' and their 'qualities' or variables (Fig. 1). Today we see that the classification is just of physical states of matter: solids, liquids and gases (Fig. 2). Again, according to classical thinking all materials could

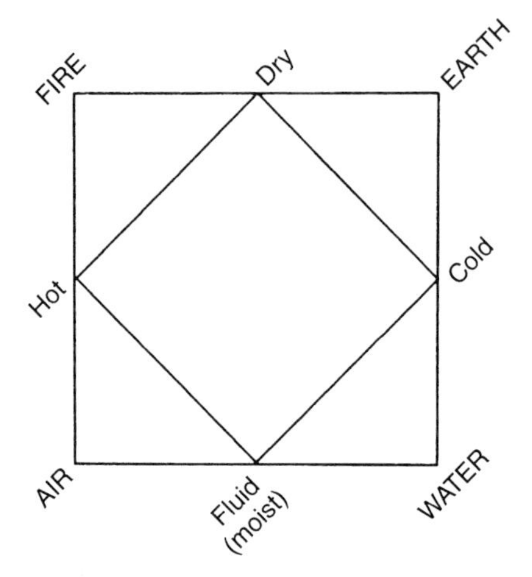

FIG. 1. The 'elements' of composition according to the Greeks, showing also their qualities or variables.

be transmuted into one another by application of the variables, qualities. For example, it was supposed that lead could be changed to gold by the action of heat. It took 2000 years of experiments to show that this classical reductionism is wrongly based. Classical thought was also incorrect in that it considered fire (energy) to be a further form of matter into which all materials could be transformed and from which all materials could be made. We have resolved their

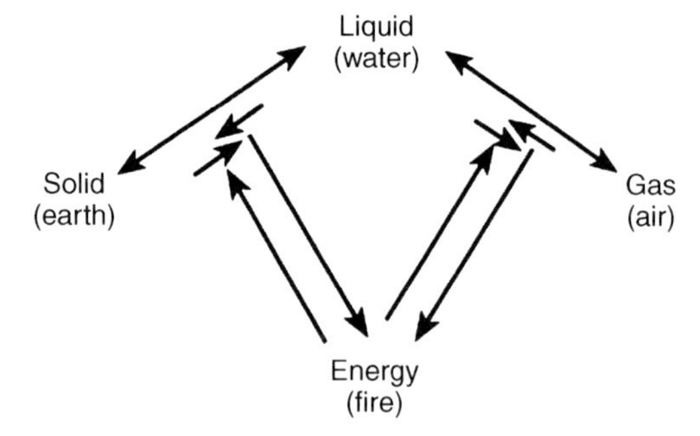

FIG. 2. A modern interpretation of the Greek 'elements' showing that they are in fact only properties of state of the elements (see Fig. 3), and that the variable is energy (fire) which has to be further analysed.

difficulties through chemical and physical reductive analysis in the last 200 years (see Williams & Frausto da Silva 1996).

The basic units of materials

Today we know that the basic units of all materials are the chemical elements. Their kinds and numbers are in part strictly limited by the properties of protons and neutrons in the nuclei and by the properties of the surrounding electrons in atoms. It is the combination of protons and neutrons which gives the limited number of elements — about 90 on Earth and some 20 more very rare unstable ones mostly made recently by man — all of which can be arranged in the periodic table (Fig. 3). As I stated at the beginning of this article, in this book we will not worry as to whether or not these three particles are open to further reductive analysis since this involves us with concepts which are not easily conceived. We can state firmly without fear of revision, however, that the periodic table gives us a final reductionist alphabet for the formation of all material around us. We shall not concern ourselves either with the steps of the formation of atoms — nuclear fusion or fission — even though we know that the abundance of the elements on Earth or in the Universe is a kinetic property of these reactions, the formation of nuclei, in giant stars. Thus one limitation on our ability to vary the construction of materials from chemical elements we recognize in passing is that imposed by these abundances of the different elements. The variable in question is that of the composition of materials, and abundance is not a very grave restriction upon it.

The variables of internal energy

We must ask next why it is that the elements fall into the pattern of the periodic table. The answer today requires us to explain the steady states of electrons in atoms. We start by posing a different question. If negatively charged electrons and positively charged nuclei attract one another through electrostatic forces why doesn't the system collapse? Initially the answer given was derived from consideration of planetary systems where collapse due to gravity is prevented by the directed circular motion of a planet round the sun in an orbit. The angular velocity is associated with an acceleration outwards, a 'force', which balances the inward acceleration due to the electrical force. The outward 'force' is called the Coriolis force. The orbit is stationary. In early models of the atom the electron, of given mass, charge and radial velocity, was then given an electrostatic attraction and an outward acceleration in balance. Unfortunately, this treatment needed an extra assumption, quantization of momentum, since it was found that electrons in atoms had only certain allowed orbits. A final working model to account for the properties of atoms and the structure of the periodic table was

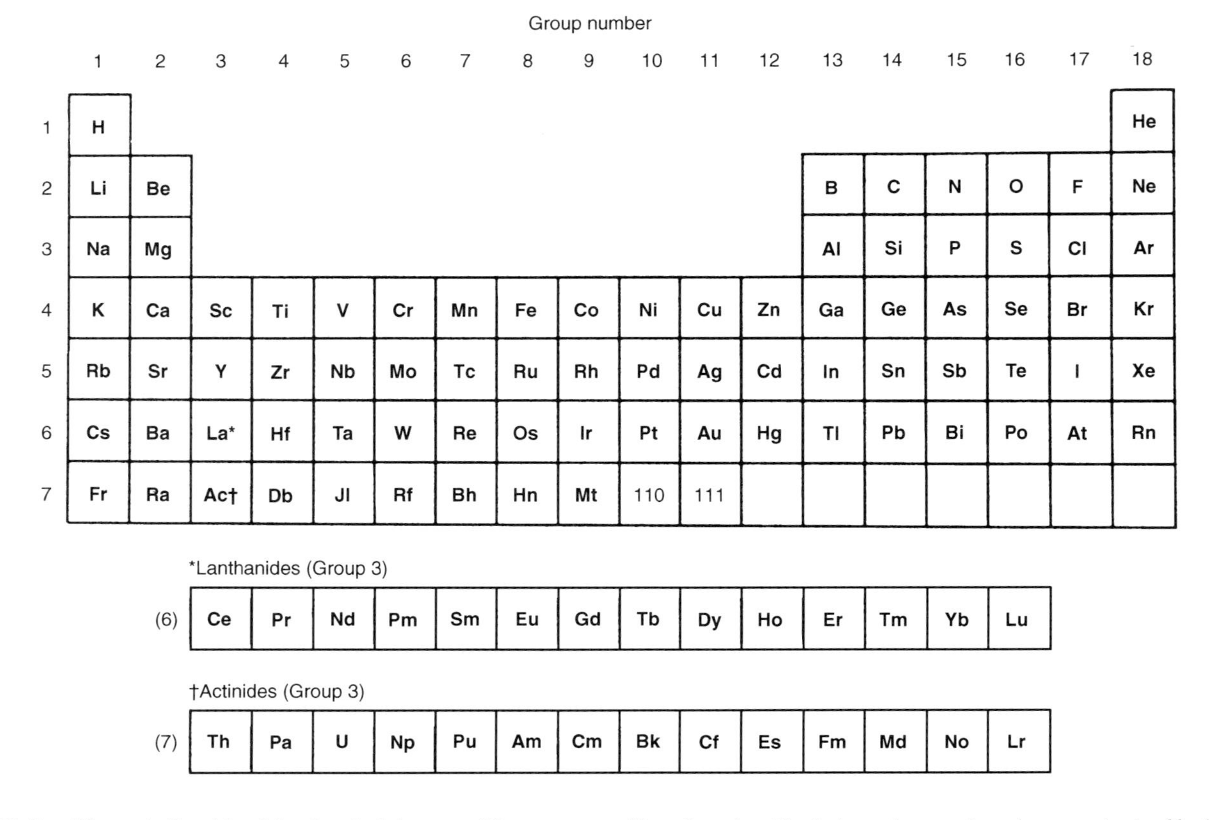

FIG. 3. The periodic table of the chemical elements. Elements up to 92 are found on Earth; later elements have been synthesized by humans. The chemical elements are the reductionist units of composition.

given by the wave equation after it had been shown that electrons (in fact all particles) behaved as waves as well as particles with mass and charge. At the same time it was found that electrons with different spin in pairs exclude other electrons from their space. Let us accept this, much though it depends on a *time-dependent* wave function. Electronic structure in atoms is then built up into shells through the limitations of orthogonal wave functions. The permitted maximum number of electrons in a shell follows the order 2, 8, 18, 32 electrons. Energy calculations now show that the final pattern of shells of the Periodic Table is 2, 8, 8, 18, 18, 32 and the beginnings of another shell of 32 before the nuclear stability limits the periodic table of elements to 92 (finishing at uranium). This is exactly the periodic pattern chemists had completed by 1900 based on the chemical and physical parallels and differences between elements. Thus empirical observations of chemists and theoretical calculations by physicists have led to the remarkable reductive conclusion that we know how all materials are made at a fundamental atomic level. There is nothing more to be discovered about the units of dead or living material. Note that in this description of electrons and nuclei, the fundamental reductive particles of atoms, we have introduced two variables: the first is the composition of the system, its atomic content, and the second is the particular arrangement of electrons around nuclei which is a function of their quantized potential energy and their radial and directed-angular kinetic energies. It is this quantized energy variable which in part also controls the existence of associations of atoms in materials.

Before we turn to these materials and their further reductive analysis we must take a quick look at what lies behind the restrictions on the variable energies of electrons. The wave theory of quantum mechanics puts together in one equation (1) waves, mass, charge and space

$$\frac{h^{-2}}{2m}\frac{d^2\psi}{dx^2} + V\psi = E\psi \tag{1}$$

where x is the space coordinate, m the mass, V the potential energy, E the total energy and the first term in the wave function ψ is the kinetic energy. As a wave, an electron stretches through all space to infinity. In fact the equation removes our ability to *fully understand* what is implied by the idea of reduction in terms of particles or energy. The symbolic logic we use no longer matches any conceptions we have and the wave equation is a postulate not a deduction. Is this a failure of reductionism? Having mentioned the difficulty that reductionism seems to collapse at this level into abstract metaphysical mathematics I return to the association of atoms in materials around us.

As an example we take the problem of the reductive analysis of the molecule of water, H_2O. Given the wave equation we can explain its particular stoichiometry, its molecular shape and virtually all its molecular physical and chemical properties. Truly a triumph of reductionism. Thus the evaluation of the most stable combination of the variable potential energy and kinetic energy of limited motion (suitably quantized) of electrons in atoms has given a full description of the way the atoms H and O can come together in H_2O or any other combination of H with O. Selective energetic advantage is shown to be with molecular H_2O over all other combinations of H_nO or HO_n. Thus we can say that we fully understand by reductive analysis why H_2O is formed and we know that given the appropriate circumstances of temperature and pressure (see below) H_2O will reform unaided from H and O. The result of mixing H and O is to give H_2 and H_2O or O_2 and H_2O in amounts depending on the setting of the variable composition and consequential on the variables of binding energy. But what of the known changes of physical state of water with what we call temperature and pressure (see Denbigh 1954)?

Additional statistical variables

A variable of any system is associated with change that is either its ability to do work or our ability to work on it. The capability of doing work (or having work done on it) of a system of electrons and nuclei in atoms has so far been limited to the discussion of a single stable H_2O *molecule*. Now H_2O exists in three forms on Earth — ice, liquid water and steam (gaseous water vapour). We need variables which, over a range of energy (heat) input, describe these possible conditions. What are they? One of them we know is related to the classical idea of fire and today is co-related with two properties (variables) of molecules or atoms in *bulk* materials not single molecules: (1) *random* motion and an associated random kinetic energy of an ensemble of particles; and (2) random direction of emission of radiation, black-body radiation, again of an ensemble of particles. These concepts are not reducible to the properties of single atoms but are statistical properties of large assemblies. They are characterized by the frequency distribution of quantized states of atoms, electronic and translational, vibrational and rotational motions. We describe the probability functions of these ensemble occupations by the variable, temperature. This probability function is related to part of the so-called entropy change which, if it is to be increased as recorded by increase of temperature, requires an energy input. Thus it is seen to oppose the potential energy of coming together of particles into condensed states in which random translational motion is lost.

The second variable of state of an ensemble, again not related to single atom properties, is the volume of the system. This is related to the probability of occupation of spatial coordinates. If expressed in terms of a fixed number of atoms (usually a gram mole, 10^{23} atoms) then it increases as a function of volume or with reduction of internal pressure. The probability function includes all configurations of molecular (conformational) states. It is again called an entropy and is related to an energy that increases with volume. Obviously, increase in volume (or decrease of pressure) opposes the coming together of atoms due to potential energy terms. Returning to water, we say that at low temperature and high external pressure the binding energy of water molecules to one another is larger (energetically) than the energy required to generate a vapour equal to the pressure, that is the entropy under these conditions is small, so that water overwhelmingly exists as ice. Raising the temperature and lowering the pressure of ice automatically leads through melting to liquid water and then to boiling to water vapour, when the vapour pressure equals the external pressure. Such equilibrium thermodynamic analyses of variables are applicable to all stationary systems of atoms and molecules in physical and chemical exchange balance. The system is fully described by energy variables at constant pressure subsumed under the symbol, ΔH, potential and fixed radial and angular kinetic energy of electrons in atoms or atomic combinations and entropy terms subsumed under the symbol, $T\Delta S$. We write the capability of doing work as ΔG of any stationary state at constant pressure, as

$$\Delta G = \Delta H - T\Delta S$$

When a system is in balance — not capable of change — then $\Delta G = 0$.

A reductionist can claim that all properties of matter for any one system in bulk have been reduced to an examination of the (statistical) properties of matter, where matter is effectively in atomic units and the variables are composition, potential and fixed kinetic energy, temperature and pressure (Table 1). The fixing of all these properties and the total amount of material present decides even the shape of an object at equilibrium. Where there are many separated systems divided by *compartmental* barriers and each has its own equilibrium ΔG, this ΔG now includes the effects of bulk fields (bulk potential energies) due to interaction between all the compartments present. It is again possible to describe these stationary systems completely. We must all agree that this reductive analysis is a remarkable achievement. *It is applicable to all isolated molecules or assemblies of biological molecules in equilibrium. Reduction to properties of single units is impossible.*

Systems undergoing change

At this point we can declare that reductionism is highly successful provided we ignore the loss of the ability to fully explain anything at the atomic level in conceptual terms — only in mathematical equations. We turn then to the problems of systems which are actively changing in flow (Williams & Frausto da

TABLE 1 Limits of reductive analysis at this symposium

(a) Materials are reduced to atoms containing electrons and protons.

(b) Attractive static potential energies are reduced to mass, charge and distance of separation.

(c) Repulsive static potential energies are reduced to charge and distance of separation but for electrons in atoms include spatial exclusion (Pauli Principle).

(d) Stationary directional dynamic forces are reduced to changes of directed momentum with fixed angular momentum, acceleration or deceleration.

(e) Temperature is reduced to the fixed statistical analysis of occupation of potential energy states or of random kinetic energies of all kinds of random motion, i.e. translation, rotation or vibration. It relates to the entropy of a state.

(f) Pressure is reduced to the fixed statistical analysis of occupation of volume elements (a second part of entropy). Here we included all forms of configurational occupation of space.

(g) Directed dynamics or vectorial flow of energy or of material, generates change in any or all of the above and cannot be treated simply. Here rate of change as well as flow and their controls concern us. They generate planned activity but require sources and sinks. The concept of time also has to be analysed as well as the notions under (a) to (f). It is very questionable whether a truly reductive analysis can be achieved (see text).

Silva 1996). The example we use is again water. What is the difference between a lake and a river? A lake can be fully reductively described by the above procedure. A river flows and has an energy not yet described, i.e. the kinetic energy of unidirectional motion. Reductionism leads us to define the water as molecules in the river as above, but the energy of flow is now an applied external energy. The system demands us to define the influence of the Sun, the external energy source, on the water system since it raises the water in the gravitational field continuously before it returns to flow in a bed. But a river is not rain. To define the river we must include the contours of the land through which it flows since this gives the rate of flow and the loss of potential gravitational energy returned as heat to the atmosphere (e.g. as shown to be true at waterfalls). In other words energy is dissipated in random (turbulent) motion. What is more a river develops with time since it erodes the contours of the land continuously. One river is different from every other river and every river is different every day. What can a reductionist say? Nothing? Any one river is *fully* understood only as a holistic concept of the total of environmental effects on the water in the river. What is the problem with reductive analysis? What is not common between a river and all those other chemical and physical bodies which we reduce to atoms and the variables, composition, potential and fixed oscillating kinetic energies, temperature, pressure, amounts and fields? The answer must be that a river is described by a *cooperative set of sequential operations* done continuously on water, and continuously changing. They are sequential in time and space. Consider other things around us to which this description applies.

When we described the formation of stationary molecules or physical states of H_2O we did not need to say whether we started in any order, say from 2H atoms and added O. Nor did we ask which atom should be given which spatial coordinates at a given time. The molecule fell into place because the reductive components $2H + O$ had a best time-independent arrangement inherently derived from the components independent of starting conditions. The system was fully reductively explained without reference to external aids for reforming it from parts. This is true of all stationary conditions of atoms if we allow also certain trapped 'good' but not necessarily 'best' energy arrangements, defined by the same atomic units but increasing the variable, energy content. By way of contrast a construct which is brought about by an external agency cannot be *fully understood* from its components. Thus any construction which derives from a sequential set of operations done on it by external agencies is not open to reductive analysis since the set of operations is of necessity holistic. Thus a particular house is not a consequence of the properties of bricks but of a particular sequence (almost arbitrary!) of building. In this case the sequence of operations is planned, and a plan has a purpose. The words 'plan' and 'purpose' cannot be reductively decomposed. If we agree then we must accept that no purposely designed activity or its product can be reductively analysed. Hence designed dynamic as well as static systems are holistic. This description must include all machines. A bicycle is not a sum of its parts but a sequential (in space) assembly of parts which do not arrange themselves without outside aid. A river and a bicycle have then much in common which is not common to still water or still metal rods. It is of no consequence that the plan of the river is accidental in the sense that the folds of the Earth, even the formation of Earth and Sun were all accidental: the river remains a consequence of a multi-factorial sequential set of activities.

I conclude with reference to biological reductionism in this symposium. Man's purposeful actions are planned or programmed; geological activity is accidentally planned; biological activities became accidentally planned (by the synthesis of DNA)—every such activity is holistic to an individual construct. How can a purposeful activity be other than holistic? My arguments here in effect dismiss reductionism by *reductio ad absurdum*. To show the absurdity to which reductionism leads, I quote from Peter Atkins (1995) describing his fundamentalist 'belief' in reductionism:

> 'I see that we shall need *to build a model* of what was before there was time in a place not a space and to explore whether its consequence was *creation*. We shall, in a sense, need to model nothing, and to see if its *consequences* are this world. I don't regard that as impossible or ludicrous; I regard it as the next (reductive) logical step for the development of science'.

[I have added the word (reductive) and the italics are mine to show the connection to holism.] By way of contrast, I quote Karl Popper (1992):

> 'Poetry and science have the same origin in myths'.

Is a poem or a myth open to reductionist analysis? I take it that a myth is a story (concerning imaginary features, e.g. the wave equation?) relating to natural phenomena. Both myths and poems are sequentially designed and can only be taken as a whole. Must the same be said of science in the end since it too is the study of a continuously changing system?

References

Atkins PW 1995 The limitless power of science. In: Cornwell J (ed) Nature's imagination. Oxford University Press, Oxford, p 122–132

Denbigh KG 1954 The principles of chemical equilibria. Cambridge University Press, Cambridge

Penrose R 1995 Must mathematical physics be reductionist? In: Cornwell J (ed) Nature's imagination. Oxford University Press, Oxford, p 12–26

Popper K 1992 In search of a better world: lectures and essays from 30 years. Routledge, London

Williams RJP, Frausto da Silva JRR 1996 The natural selection of the chemical elements. Oxford University Press, Oxford

DISCUSSION

Maynard Smith: I don't think I can go along with the view that biologists cannot explain purpose and function. Most biologists do think they have an explanation of why organisms have functions. It certainly isn't an explanation in terms of saying that the matter in the organism is at equilibrium, but I'm not sure why you rule it out as a reductive explanation, and want to call it holistic if we have what we believe to be a mechanistic explanation of why a function is so.

Williams: Take for instance the observation that birds fly: this doesn't give you any explanation as to why there are birds, it just tells you that birds fly.

Maynard Smith: Perhaps biologists are simply wrong, but we believe that for 150 years we have had an explanation of why that particular functional nature of birds evolved, and it is in terms of the dynamics of populations of replicating entities.

Williams: But in this case you don't come down to the level I am talking about, which is the level of the molecules that function in the system: what you are stuck with is a global explanation.

Maynard Smith: Well, it is not an explanation in terms of equilibrium thermodynamics, that is for sure. Rather, it is in terms of the populations of replicating entities. Given populations of replicating entities you will get purpose and function.

Williams: How do you start such a system?

Maynard Smith: That's a question that goes back to one of the issues Thomas Nagel raised this morning: is it possible to understand how replicating entities could have emerged on the surface of the earth without somebody coming along, putting them together and letting them go? At the moment we cannot give a full explanation of the origin of replicating entities (although I think we're probably not too far away from this). This is the issue: if indeed it is not plausible for chemical replicating entities to have emerged on the surface of the Earth, given the conditions at that time, and that some other force had to put them together, then one would have to conclude that ontological reductionism (as Thomas Nagel puts it) is correct. At the moment we are a long way away from reaching that conclusion: the evidence currently leans in the direction of thinking that we will be able to provide an explanation for the transition from chemistry to biology.

Williams: Even if the process is accidental?

Maynard Smith: If it's the kind of accident that would have taken millions of universes millions of years, then I think we are faced with a genuine ontological problem. If it is the kind of accident that is going to come off as soon as you set it up in the test tube, then we are not.

Nagel: Bob, is it your view that the origin of life is antecedently overwhelmingly unlikely?

Williams: Given the physics and chemistry as I understand it, my view is that it is extremely unlikely. As we see it, it is a one-off event and as such, in my view, cannot be given a scientific explanation. If it can be set up in a test tube *regularly* it can be investigated, of course.

Kerszberg: The concept of the probable versus the improbable is a difficult notion to manipulate. For example, even if life is an extremely improbable event, if the universe is infinite (which cosmology doesn't rule out at the present time) then life must not have occurred only once, but an infinite number of times despite the fact that it is improbable. Things happening at a certain level of description, sometimes because of strange numerical coincidences, can conspire to create the possibility for improbable accidents at different levels. You talked about angular momentum. This is a particular case of what Freeman Dyson has called 'cosmological hang-ups'. When things are turning they cannot fall on each other: planets do not fall on the sun because of the conservation of angular momentum. Because of that, general conditions have remained the same on the Earth for a very long period, making extremely improbable things (such as life?) possible after all.

Mitchison: The last time I was in this room was at a meeting on *The origins of life in the solar system* (Ciba Foundation Discussion Meeting, 27 Jan 1994), and one striking fact which emerged was that initially the earth was essentially uninhabitable because it was being bombarded. Late heavy bombardment ended about 3.8 Ga ago (McKay 1991) and then life appeared more or less immediately (the first stromatolites are dated $\sim$3.5 Ga; Walter 1983). From a Bayesian viewpoint this would argue that life was just waiting to make its appearance, thus that it was highly probable.

Williams: That is a very reasonable point of view. The other proposition is that it just happened.

Gray: Of all the planets where life might have evolved, we have to be living on one of them where it has happened. This is the so-called anthroic principle. It is therefore odd to say that the probability of that accident having taken place is enormously low.

Rose: I actually don't think that the origin of life is particularly improbable. But, as I understand Bob's argument, it's not at all dependent on the question of the probability of life: it is dependent on the question of whether what he calls purpose and John calls function is implicit in and can be read-off from the molecular or structural compositions of the components. Manifestly, it can't, because you need the historical dimension which comes in the moment you begin to talk about biology. It seems to me that the problem is that we're mixing various traditional philosophical categories of concepts of causal and functional explanations in biology with the several different Aristotelian definitions of cause. Unless we are prepared to agree that there are many different appropriate ways of explaining biological phenomena, we're going to get stuck. In this context I was struck by something that Thomas Nagel said about not everything existing with the kind of reality that science is after. Embedded in that is a concept that the only legitimate type of explanation is one that actually comes from the reductive elemental and fundamental composition of things. There are types of explanation within biology which accept the reality of purpose/design/function but can't simply be read-off with what you regard as the only legitimate scientific rational reality.

Nagel: It is an open question as to whether we can expect to stop in our search for the real explanation of biological phenomena at the biological level.

Rose: Why is there just one real explanation? Why can't we have multiple legitimate explanations of the same phenomenon both in terms of molecular composition and in terms of its functional and causal roles?

Nagel: We certainly can, but if the functional explanation has no independent causal reality and is simply the manifestation of something that is derivable from physics, then it is explanatory only in the sense that it gives us a kind of understanding, but it doesn't tell us about how the real world works.

By the way, I don't think that everything is scientifically explainable. There are many things such as questions of value that simply don't come within the remit of science, but the natural order is the subject matter of science — why things happen. There is a real question as to whether biological explanations — which of course are essential to give us the understanding, comprehension and predictions that we want — are causally fundamental.

Bateson: One part of Thomas Nagel's distinction emphasizes what is it we want to know. If we ask a question about function, we don't necessarily want to know about history. Take the example of a blue tit learning to pierce through the foil of a milk bottle. Here the historical explanation does not involve evolution by natural selection. A bird has discovered how to obtain cream by chance and then this behaviour is transmitted to other individuals culturally. We can ask a question about function and be satisfied by the answer: the blue tit pierces the milk bottle top in order to get food which may make the difference between survival and death in a cold winter. A functional explanation about current utility does not imply anything about the historical origins of that behaviour. That is why it is helpful to specify what it is we want to know.

Quinn: I'm not sure that understanding the function of a bird's wing from pieces is as helpless as understanding the function of a bicycle or steam engine from pieces, because bicycles were invented by someone with an idea and birds presumably evolved with maximum functionality that isn't very far removed from other local maxima. So it's possible to dissociate the problem of the actin and myosin generating motion in the muscles from the separate issue of how to make bones light enough. Reductionism in biology often works much better than one would think from the physics or chemistry of machines.

Dover: With regards to machines and their functions, you make the valid point that in biological evolution the bicycle (or its biological equivalent) is put together by accident rather than purpose, yet you still have this image that there is this object called a bicycle (or its equivalent) which is a functionally equivalent piece of machinery. I think that's a false way of looking at biological evolution, because evolution could have taken all sorts of other directions. Many other combinatorial permutations of 'machine' parts might have taken place which would have given rise to functions, which would not be the end state we call a bicycle but which might, nevertheless, successfully function. If one looks at the whole of 'phenotypic space', the space of functions in biology, the vast majority of this is empty. This is not because this space represents useless functions that were never allowed to progress, but simply because, historically, it has never been filled. The bits of space that happen to be filled represent particular lineages of accidents that have produced the limited number of functions we see today. My contention is that things could have happened differently: there could have been different starting points and other bifurcating lineages taking place which would have led to different yet successful functions. Therefore we shouldn't be seeing the bicycle as the only efficient way of carrying out a particular function: many other bizarre and

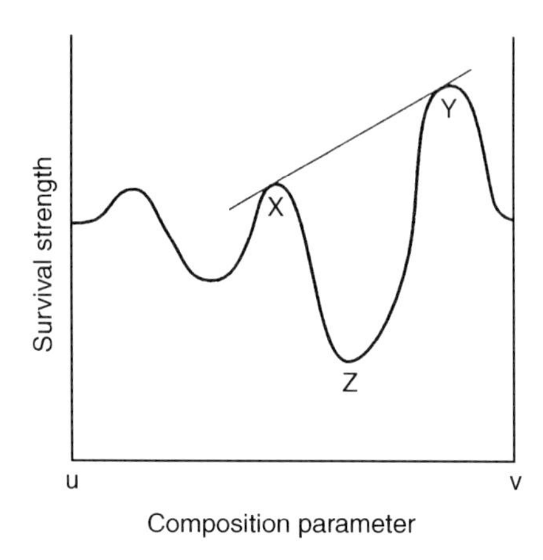

FIG. 1. (*Williams*) Survival strength is plotted for organisms that, because of their internal (hidden) systems, could exist at a given composition on the x-axis, u–v. Consider a situation in which two organisms of composition X and Y grow in the same bath containing components u and v. As Y, the more favoured, grows it removes more of the components v, and therefore pushes the composition of the bath so as to favour X. Organisms between X and Z and Y and Z cannot survive since the combination of X + Y always outperforms them. Such figures inevitably result if the interactions in the kinetic and thermodynamic schemes within life are differently cooperative for different ratios of components u and v. A plot of survival value against some property—here, chemical composition—is often called a fitness landscape of species. In life, the number of chemical elements is 15–20 and the resulting components, u and v, become very large indeed, so that the figure has a large number of dimensions. The particular species which survive are not fully determined and the system is not open to reduction in my terminology.

seemingly 'unnatural' biological functions might have arisen and all of them would be inefficient because they are all products of historical accidents. Hence, we should not be asking questions about the probability of producing a bird's wing from its constituent 'machine' parts. That is equivalent to asking the largely uninteresting and irrelevant question about a given person winning the lottery. The process of evolution is such that new functions can arise, willy-nilly, no matter how messy or inefficient, within the obvious constraints of contingency. Hence, the probability of a function arising is one, in the same way that the probability of someone winning the lottery is one, once the wheels are turning.

Williams: I will try to illustrate my point using a landscape diagram (Fig. 1 [*Williams*]). If you plot survival on the y-axis, you can ask what the variables should be along the x-axis. The living machines that exist are around the places of high survival strength, X and Y, and the machines that don't exist are of low survival strength, Z. When you explore the whole of space on many different

x-axes, i.e. using different variables such as chemical element variability, energy intake and temperature, what you expect is a series of peaks and troughs. The real situation has many dimensions and survival may be the result of a curious accident of a combination of variables: not the best, but a number of good answers of high survival value which co-exist. That is like a turbulent situation of stellar particles evolving into a number of galaxies, which is inexplicable by reduction. Evolution could have happened in a very different way to find different peaks.

Henderson: In terms of its effect on life, is the relationship between the levels we're discussing really so critical? If you were to change the physical forces such as the magnitude of the electronic charge or Planck's constant, how much would the periodic table actually change? You argue that the numbers 8, 8, 18 are important: those are not affected by small changes in the attractions between nuclei and electrons and so on. In chemistry, how would the periodic table be changed if you were to alter the forces slightly? How would that affect the chemistry of the atoms, and at what point would water not form? This gives you some idea of how robust the path back to particle physics is.

Williams: The periodic table is based on space–time considerations of waves. If you are defining wave/particle motion around a point (here electrons moving close to a molecule), then you build up angular momentum in shells in one way only. For zero angular momentum, the first wave is spherical and of wavelength $\lambda/2$. You are allowed two states for each wave, which we don't really understand, called the states of spin. Thus the first number of waves, called $s=1$, is 1×2, with zero angular momentum. The two form the $1s$ shell. The next spherical wave, $2s$, is wavelength $3\lambda/2$, and again there are 2. If you now take an angular momentum of one, which has a wave of wavelength λ' which goes round the nucleus, the ways in which you can express those which are orthogonal to one another is three: since each has two spin states, that is $3 \times 2 = 6$, which is part of a period after $2s$ called $2p$, so the total for the second period, or shell, is $2 + 6 = 8$. If you double the angular wavelength, $2\lambda'$, the number of orthogonal functions you find is five: $5 \times 2 = 10$, and they give the transitional elements, belonging to the $3d$ subshell. Now there are $6(3p)$ and $2(3s)$. The total number of electrons in the third period is then 18, i.e. $2 + 6 + 10$. This beautiful series therefore falls out from properties of waves not particles, and the periodic table is a fantastic and unique feature of space and time from which all atoms are created. Reduction to this level is perfect. However, if you now destroy the relationship that gave these solutions in some way then we would have no idea what we are talking about without defining the mathematical system all over again. It creates a terrible mess if you start fiddling with constants such as Planck's constant or units of charge. They are invariable. A final point in my text is the peculiarity that by using the wave equation for an electron we actually lose the idea of mass points moving in space and time and we have reduced to uncertainty, not knowledge in the old-fashioned sense.

Ashmore: I want to go back to a concrete question which you raised in your talk, the whole issue of protein folding. To me it seems that this is the major dividing line between the life and physical sciences. As far as I understand it, the protein folding problem is unsolved.

Williams: You are talking about the mechanism of folding, not the final form. We believe that the final form of the vast majority of proteins is the thermodynamic ground state. Proteins then fall into the category of components as chemicals of a system, but their melting is a complex problem, not one reducible to linearly related variables.

Ashmore: Predicting protein folding may just be a computational problem, but the computational problem is far beyond what we're actually capable of doing.

Perutz: The problem is not beyond capacity for smaller proteins. The difficulty is not really one of computing: rather, the interatomic forces between unbonded atoms are not sufficiently accurately known to produce a unique answer.

Bray: Biologists think that chemistry is an exact science that is well defined and firmly based in physics. Bob Williams, you started off talking about particles and the forces between them, and wave functions, and then you were able to describe a hydrogen molecule. You then went to water and glucose. How far up can you carry on in that way, being firmly based in particle physics? What happens after that? For instance, what sorts of explanation are you using when you talk about phosphorylation of a protein?

Williams: We would never dream of trying to use wave functions to describe this. If you deal with static chemicals which are relatively small (e.g. glucose) and don't ask anything about the dynamics (put them at 0 K) and put in the forces as we know them, you will get a very reasonable explanation of why glucose exists as a stationary state of carbon, hydrogen and oxygen. It is not a stable state relative to carbon and water. When you introduce glucose into a living thing or into a chemical reaction system, the problem becomes immense because you enter the realm of dynamics as well as that of statics. The dynamics in the first instance is to do with temperature and pressure, which are extremely difficult to handle at this level (300 K is a nasty temperature). Now, if you also introduce flow asking how a living body handles glucose, then you won't find a chemist who can solve those problems. Therefore the answer to your question is that we don't get very far in chemistry using this sort of theory. Reduction fails to be of any value. We might believe it would work, but when we get to a certain size we don't care, because the complexity of the system leads us to believe that we wouldn't be able to solve the problem at this level, and there may well be a multiplicity of possible answers only one of which we know about.

Perutz: Surely the phosphorylation of a protein can be understood reasonably well in terms of empirical laws and transition state theory. The geometry of the interaction of the protein with the substrate reduces free energy of the transition state and allows phosphorylation to take place at room temperature. What is wrong with this explanation?

Williams: You called it 'transition state *theory*', and that's what it is: it refers to a particular theoretical pathway of events during reactions. It is an approximate and useful theory which no respectable small molecule chemist believes in at all. It's a useful theory but only to a certain level: it isn't useful at all if you are seeking ultimate understanding of energy distribution in bonds leading to reactions. Such a reductionist approach becomes too complex, however, to be meaningful for enzyme kinetics and we use the elementary theory you describe.

Dover: Phosphorylation might be explicable in terms of chemistry, but the biological question concerns why that phosphorylation is taking place, not how.

Holmes: The concept of phosphorylation is an interesting example of where a certain kind of reductionism will be important. Ultimately, to understand energetics of proteins, we're going to have to try to describe the reaction coordinates of such reactions in much more detail that we can as yet.

Gray: I want to drop a few more pebbles in the pond. First, I find it odd that one could even contemplate that, if you just follow the chemistry up, you could account for the stability or otherwise of particular molecules in biological systems. Much of what biology is about is the selection of control systems that are there precisely to make things either stable or change under controlled conditions. So there has got to be a point in biological systems when you stop talking about chemistry and begin to talk about the way that machines have been constructed by natural selection to do certain things.

Second, going back to the issue of what is meant by an explanation as opposed to a description, there might be some beautifully general philosophical way of answering that question. However, in a simple-minded way, an explanation is a satisfying answer to a question — typically, a 'why?' or 'how?' question. We're all familiar with many perfectly reasonable questions which don't depart from the empirical world of the realm of science, yet to which the answer is quite clearly couched in terms of neither physics nor chemistry, nor even in terms of biology. If we return to Thomas Nagel's example of the piano player, we can ask all sorts of questions about, for instance, the way the nervous system works or the evolution of a system that can enable piano playing, but none of the answers to these questions would tell us why Brendel played an all-Beethoven programme last night. If you are going to answer a question such as this, it wouldn't even be sufficient to go back to the history of Brendel and talk about how his brain was trained and all the changes that took place because of reinforcement mechanisms, etc. No answer to that question could avoid using concepts about music and Beethoven, and those questions will not get any kind of explanation within the levels that we've been talking about. Instead, they must be answered in terms of human history, culture and sociology.

Noble: I thought that's precisely why Thomas Nagel introduced the concept of type. This is why I raised the question about the division he made between ontological and epistemological resistances to the reductionist program. If one follows the logic that Thomas Nagel was pursuing, it leads us to say there are many other things there. It is not just a matter of whether your resistance to a reductionist program is epistemological or ontological—they interact. Putting this another way, what there is empirically does depend on our concept of the world.

Morgan: Jeffrey Gray and Steven Rose have presented a democratic pluralist theory of levels. Their point seems to be that there are many different sorts of explanations which one might want in different contexts, and we should not be undemocratic and say that one level is more fundamental than another. The trouble with this position is that it is of no practical use whatsoever, because the real arguments about reductionism are actually political: that's why it's an important issue. They are to do with deciding at any one time how resources should be allocated within science, which in turn consists of deciding what 'good science' is. For example, take the brain. If you want to understand how the brain can perceive, how are you going to set about this as a funding body? If you're very adventurous, you give lots of money to me—I'm a psychologist—and I'll give you an answer. If you are more sensible, you might give the money to Horace Barlow, who is looking at properties of single nerve cells in the visual system, or you might go to Jonathan Ashmore for some biophysics, or you might go the whole hog and give all the money to Martin Raff and the molecular biologists as some neuroscientists seem to be advocating. I'm not going to take sides on this, I'm just telling you that this is actually the position. I would argue that the democratic pluralist view is useless because it doesn't help to solve any problems. Furthermore, it allows anything to go: it allows you to say that if we want to understand visual perception, we can't do it without a feminist perspective, for instance. I feel the virtue of the hard reductionist position, which says that there is a fundamental level of explanation, is that it imposes discipline. Coming back to the fact that I'm a psychologist, I could produce thousands of models of visual perception. However, if I am required to produce something that accords with known properties of nerve cells in the visual system I immediately become extremely constrained, and I have to do a whole lot better. This is the virtue of the idea that there is something fundamental to biology. Understanding the brain at the physiological level is more fundamental than producing a purely verbal account of how it might work. And, of course, there are things more fundamental than physiology that keep the physiologists honest.

Nurse: I think the periodic table is one of the most beautiful constructs in science. To what extent is it arbitrary? Does spin have to occupy two states?

What if it can only occupy one? Are we building a construct on the basis of certain arbitrary assumptions?

Williams: I have described the nature of the units in materials as we know them today. The ancient Greeks thought that they had got the description right on a completely different and wrong-minded basis. Two thousand years from now people might say that we got it all wrong, but the internal consistency of our solution is remarkable and useful. If you talk to Penrose, who is a fine mathematician, he will say there are parts of physics that are so little understood that we don't know where we are in this world, for example, in the study of the brain, and that there could be ways of giving explanations other than the ones we accept today. I have used wave mechanics to describe atoms, and when I use wave mechanics I use an equation in which there are things we really know little about. We use words such as mass and charge to describe forces but we know very little about what these are. Similarly, spin is difficult to describe again, except by mathematics, at which level classical science seems to collapse. Since the middle of the last century, we have lost material images as the basis of reductionism. When we have to use mathematics we meet a semi-mystical metaphysics! This is where science-based reductionism really finishes. However, using these 'concepts' we can say with a very high degree of certainty that the limits of chemistry are known. For myself, I consider the periodic table to be 'explained' down to the level of wave mechanics.

Kerszberg: We enjoy the luxury of discussing reductionism because we have built an understanding at various levels and we can now examine the relationship between these levels. Had we had to wait until we knew everything about a 'fundamental' level before we could proceed to understand the next level and so on, we would be nowhere. When you say that mathematics is the endpoint of reductionism, even that is not true because the mathematicians have their own problem — the Gödel incompleteness theorem — which tells them that they will never be able to build a closed coherent formal system. But this doesn't prevent us from using mathematics in everyday life and even in science. It is fortunate (but no accident) that the incoherences of one level of description seem to become irrelevant at the next level, and we can still use all sorts of useful facts from one level at the next. We don't have to wait until the particle physicists have come up with their theory for everything.

Garcia-Bellido: When one looks the different levels of organization, going from quarks to populations of individuals, there is a clear increase in complexity. It seems that there must be some underlying trend in the necessity of elements to combine; some logic to this kind of increasing complication. If we analyse any particular level, it becomes clear that this level cannot be explained in terms of the elements underneath it, because surprisingly enough these elements have been understood by the way they interact in order to explain the level that we are observing. So it is not a property of the element of a level of complexity by itself. Therefore there is a fallacy in the reductionist principle that if we go all the way down we are going to find properties which in themselves are directly observable and once we measure them we're going to combine them at the next level up, and so on. This explains why we have to work out the levels as they are. We must explain them not in terms of the ultimate level of reduction, but in terms of the level which is exactly underneath. Once we do this, we are surprised by the fact that there's no way that we can predict the combination at the upper level: for example, that the DNA is made out of just four types of nucleotides out of the very many that exist. There may be some physical law which we will discover one day that will tell us that there must be four nucleotides making DNA, but it seems that they are those that amplified and generated the next level; this is where the contingency starts appearing. This is also true for the sugars, the optical rotation or the amino acids. Surely it is an element of contingency, but it doesn't mean that the contingency is random. Usually contingency is between very few degrees of freedom. That means there are internal possibilities for solution which are not unique, but while they are there the system is fixed and they can only simply amplify. Moreover, any level is not constructed with all the elements of the previous one, but only with a sample of them: the others are simply left at that level and will not play a role at the next one. Increasing complexity is merely sampling them, combining them and multiplying them, making a new rule of the game. This new rule can only be ascertained by looking to its own level of complexity. There is no way we can leap from nucleotides or DNA to Mendel's laws. It looks as if, although there are degrees of freedom, these increase with the levels of complexity. Increasing complexity leads to increasingly more ways of solving equations or solutions. Now we don't know the transformation rules because perhaps we are putting too much emphasis on trying to reduce the phenomena to things which are at much lower levels, and we are losing the very power by which we can understand them. It is silly to start thinking that the way to understand development is by looking at the structure of the gene or even 100 000 genes: the way they are going to interact is not written in the structure of the genes. I think it is a mistake to leave this complexity to be explained in terms of the properties of the elements.

Nagel: If there's an explanation of the behaviour of a complex system, then there are three possibilities. One, which I think holds in the case of why Brendel played an all Beethoven programme last night, is that this is an explanation in its own right that cannot be reduced to anything else, and which really explains the thing that happened. Another possibility is that it is an explanation that is correct and that can be reduced, just as we can reduce the periodic table to atomic physics. There is a third possibility, which I tried to illustrate in the case of artefacts. For example, the functional explanation of what a typewriter does gives us an understanding of it, but it isn't really what's happening. Of course, it can be explained by human psychology which designed typewriters. But I do think there is a question about biology, whether it falls in the first category or the third.

References

McKay CP 1991 Urey prize lecture: planetary evolution and the origin of life. Icarus 91:93–100

Walter MR 1983 Archaean stromatolites. In: Schopf JW (ed) Earth's earliest biosphere, its origin and evolution. Princeton University Press, Princeton, NJ, p 187–213

Macromolecular structure and self-assembly

Richard Henderson

MRC Laboratory of Molecular Biology, Hills Road, Cambridge CB2 2QH, UK

Abstract. The output from the molecular biology revolution has grown steadily and logarithmically from the first protein sequence, insulin (Ryle AP et al 1955 Biochem J 60:541–556), the first three-dimensional atomic structure of a macromolecule, myoglobin (Kendrew JC et al 1960 Nature 185:422–427), the first DNA gene sequence, ϕX174 gene J (Sanger F et al 1977 Nature 265:687–695) and the first genome sequence for a free-living organism, *Haemophilus influenzae* (Fleischmann RD et al 1995 Science 269:496–512) to the current situation where the output rate is close to one new gene sequence every few minutes, several new three-dimensional structures a day and a new (bacterial) genome completed every few months. Those working in this field must readjust their horizons to this changing situation every year or two. In the area of three-dimensional structure of macromolecules and macromolecular assemblies, the methods of X-ray crystallography, nuclear magnetic resonance and electron microscopy have combined to produce powerful insights into how these molecular machines work. In this paper, I present three examples of molecular machines whose structure tells us a lot about how they work. These are the light-driven proton pump bacteriorhodopsin, the ATP synthetase molecule which contains a tiny motor and generator, and the flagellar rotary motor which provides the thrust to power physical movement of the bacterial cell. The structure itself in three-dimensional detail is thus often seen to provide the most important single insight into how things work, reducing biology to chemistry and physics. The reductionist approach in this field seems to be limited only by the accuracy by which it is possible to describe inter- and intra-molecular interactions in terms of hydrogen bonds, van der Waals interactions and electrostatic forces. At present, there is no fundamental limit in sight.

1998 The limits of reductionism in biology. Wiley, Chichester (Novartis Foundation Symposium 213) p 36–55

One of the reasons I was glad to be invited to this meeting is because I have never really asked myself the question before of what the limits of reductionism are. Although I'm interested in the philosophical issues, I decided it would be advisable if I kept my paper within the reasonably strict bounds between chemistry and cells, in other words to keep it in the area of molecules. I'll leave it to the discussion to go outside this area.

In the area of structure of biological macromolecules, what are the levels of questions we are interested in? Two most important steps missing from a reductionist view of science are those involving discovery and invention. Much of our work is necessarily concerned with discovering the different molecular components, because we don't yet know who all the actors are in biology, although it won't be long until we do because of the various genome programs underway. Once we know what all the components are and what they might do, we can then start looking for explanations and levels of answers. One of the most important levels in macromolecular assembly is illustrated by asking the question, 'What is the structure?', because just knowing the structure will often provide the answer to other questions.

The sorts of questions we are interested in relate to the size of the molecules, their behaviour in terms of aggregation or self-assembly, and then, in more detail, the secondary structure and atomic structure of the molecular components of the various biological macromolecules. Once this information is gained, we are armed with a variety of data and we can then ask for explanations and descriptions of what is going on at different levels. Once you have a structure and know what the function is you can make hypotheses without doing any more experiments and come up with an explanation that may be wrong, but which may be very satisfying. Most people would then like some proof that these hypotheses are true, and thus they go on to make various experimental predictions against which to test the hypotheses. This is as far as molecular and cell biologists want to go. Many researchers would be happy to stop at this stage: they have created a set of self-consistent experiments that fit with the structure and function, and seem to provide a reasonable level of explanation without going back into chemistry and physics too much. However, there are others who will want to go on to make models, taking these hypotheses, then writing equations and trying to test how accurately the theory explains the observations. Others will go on to attempt a full atomic simulation, using the energy functions to describe the inter-atomic interactions: the hydrogen bonds, the van der Waals forces and electrostatic forces. Once you are at this stage you can go on and on making increasingly precise calculations. If the explanation you are looking for depends on a balance between the energies of two states (e.g. an equilibrium between two conformations, or a folded versus an unfolded protein), and the energy difference between the two states is very small relative to the full summation that you are doing, it can be that you are caught out in a situation where the accuracy of measurement is insufficient. Thus however far you proceed in understanding atomic interactions in proteins, there are some people who will never be satisfied with the accuracy that is attainable. Thus my view is that different researchers looking at the biology of macromolecules are satisfied at different levels. There is an element of choice and opinion involved at these different levels.

I will illustrate these different levels using three examples of molecular assemblies from the membrane of a single bacterium, *Halobacterium halobium*. These three assemblies cover a variety of levels of actions that are of interest here. The first is a light-driven proton pump. The halobacterial membrane has arrays of a membrane protein called bacteriorhodopsin, which I have worked on for about 20 years. It is a mini-crystal, 100 molecules across, which absorbs light and pumps protons out of the cell, thus charging the membrane. The gradient of hydrogen ions is used by this organism as a primary source of energy for a variety of functions in the cell. At various points in the cell membrane this hydrogen ion gradient can be used by a second membrane machine, F_1-ATP synthetase, to make all the ATP in the cell, and the third membrane assembly I want to describe, also driven from the hydrogen ion gradient across the cell membrane, is the flagellar motor, which rotates, driving a flagellum, which then acts as a propeller to drive the bacterium along. The structure of each of these molecular machines leads almost directly to an understanding of their function in terms of their chemistry, and for many people this provides an entirely satisfactory explanation, without any need to go to any deeper level of reductionism.

Bacteriorhodopsin

Patches of crystalline protein in the *Halobacterium* membrane containing the single species of protein molecule, bacteriorhodopsin, were discovered in the early 1970s by Walther Stoeckenius and his colleagues (Oesterhelt & Stoeckenius 1971, Blaurock & Stoeckenius 1971) using freeze fracture electron microscopy, but at this time their function was unknown (Fig. 1). Here was an interesting membrane structure without a known function. However, Stoeckenius thought these structures were interesting and that they ought to be worked on, so he got a few people to do this. At this stage there was a variety of possible explanations. Bacteriorhodopsin is purple coloured because the protein molecules have a chromophore, retinal, and isolated patches are a deep purple colour, so people thought that *Halobacterium* might absorb light and use it as an energy source. Indeed, this is the function of these crystalline arrays. However, others thought they may be acting as light sensors for the bacterium which would be used to control their swimming behaviour (phototaxis). Both these explanations turned out, in a sense, to be true: later on it was discovered that there were four different genes for retinal proteins in *Halobacterium* which are closely related in amino acid sequence (30% identity). The first one forms these crystalline arrays, absorbs light and pumps protons out of the cell. However, there is also another pigment at about a tenth of the concentration of the first which is excluded from the purple patches and is mixed up with the other proteins in the membrane. This second protein, called halorhodopsin (Blanck & Oesterhelt 1987, Schobert & Lanyi 1982),

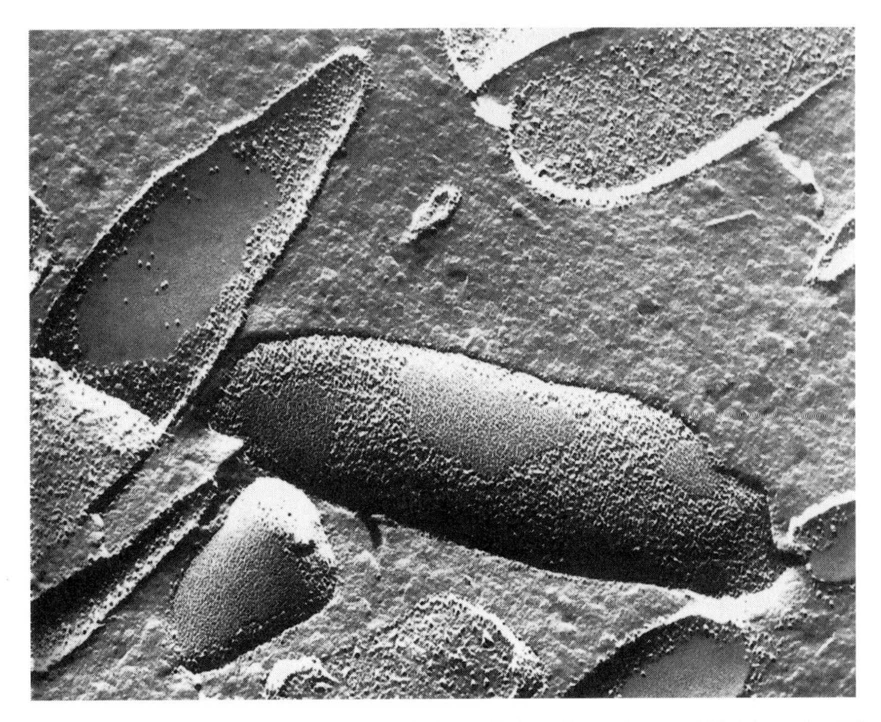

FIG. 1. Freeze fracture electron micrograph through the cell membrane of *Halobacterium*. The patches with hexagonal striations are the purple membrane patches containing the bacteriorodopsin molecule which is a light-driven proton pump. (Courtesy of Walther Stoeckenius.)

absorbs light and pumps chloride ions in the opposite direction into the cell. There are also two sensory rhodopsins, called SRI (Blanck et al 1989) and SRII (Seidel et al 1995), that are involved in phototaxis. They actually perform the function equivalent to a bacterial eye. They each have a much slower photocycle and pump no ions across the membrane, but their conformation in the light-activated form is sensed by cytoplasmic machinery, which feeds into the motile behaviour of the cell. SRI controls behaviour by attracting the bacteria into regions of higher visible light intensity whereas SRII controls behaviour by causing the bacteria to swim away from regions of higher UV light intensity. Thus all the initial guesses as to the function of this molecule were found in homologous molecules: two are involved in energy transduction and two are used as sensory pigments.

We worked on this and produced an atomic model (Fig. 2) for bacteriorhodopsin (Henderson et al 1990, Grigorieff et al 1996). The next question we asked was: knowing the atomic structure of this protein molecule,

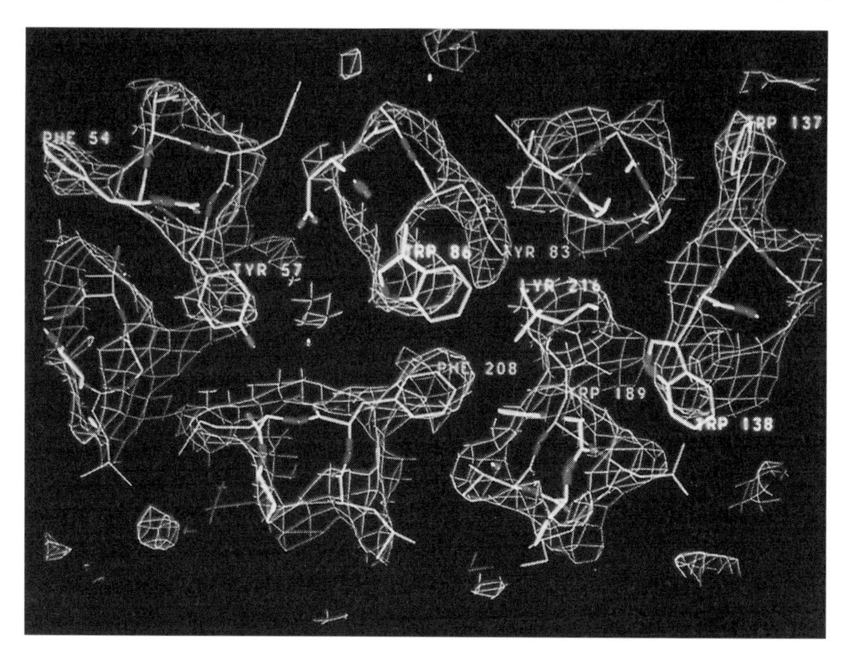

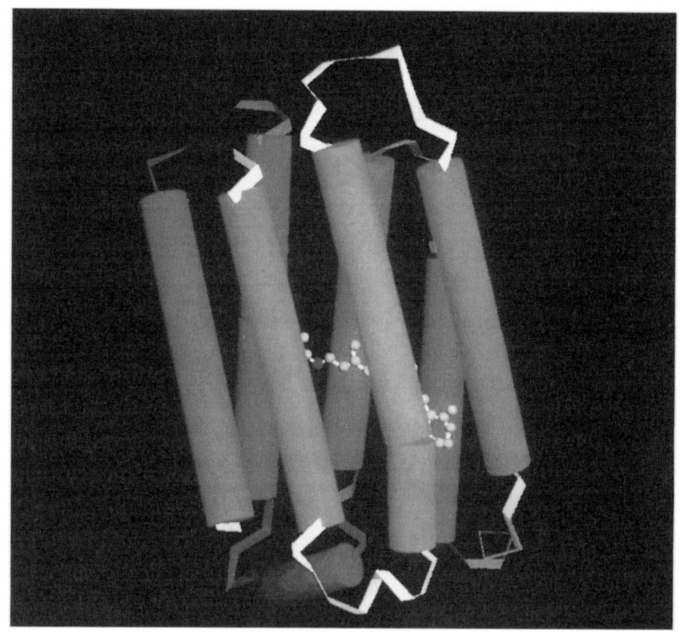

FIG. 2. (*Top*) Atomic model of part of bacteriorhodopsin obtained by electron microscopy and (*bottom*) schematic of the whole structure showing seven transmembrane α-helices.

can we understand how it absorbs light and pumps protons? There are various levels at which this question can be answered. The one we found to be the most revealing involved going to a more schematic representation of the structure rather than one showing all the atoms from which all the energy functions might be calculated (Henderson et al 1990). The purple chromophore, all-*trans* retinal, is covalently linked to one of the lysine side chains in the structure. In the centre of the protein, half way across the membrane, we find the retinal placed between two half-channels leading from the centre of the membrane to both the cytoplasmic and extracellular sides. When light is absorbed by this chromophore an isomerization takes place and a proton of the protonated Schiff base of the lysine residue (Lys216) is released via an aspartic acid (Asp85) to the outside, and 10 ms later another proton from another amino acid (Asp96) re-protonates it from the other side (Fig. 3). Thus you can understand the function of the protein molecule in terms of a single group with an affinity for the pumped proton which changes as it goes through a series of different structural states.

We have therefore obtained quite a satisfying explanation of how things work at this level (Henderson et al 1990), and for some people this is enough. But for others it is not, and Fig. 4 shows another representation of what is happening (Lanyi 1992). In this experiment the bacteriorhodopsin is initially in the dark with the chromophore in the all-*trans* state. If you flash this with a short pulse of light you can detect a series of different intermediates which all have slightly different structures. We don't know exactly what these are, but the kinetics of this photocycle has been measured extensively and the experimental measurements of the build up and decay have been simulated by a mathematical theory. The differences between this theory and what is really happening represent deficiencies in the explanation which some people are interested in and continue to work on, whereas others think we have gone far enough.

Some researchers have taken the complete list of the atomic positions and done calculations of the different energy levels of the whole structure, including a simulation for the water molecules and the dielectric constants of all the components. The question they have asked concerns whether from the atomic coordinates of the experimentally determined structure we can predict how the affinities of the different ionizing groups for the proton that is pumped behave (Sampogna & Honig 1994). Figure 5 shows the titration state of four of the Asp residues in the middle of the protein, two of which (Asp85 and Asp96) are known experimentally to be involved in proton pumping. Sampogna & Honig (1994) find a reasonable fit: the residue that is always unprotonated (Asp212) has the lowest affinity, the one most heavily involved in the early stages of the proton pumping is the next (Asp85) and the two which are normally protonated at neutral pH, waiting for the second part of the photocycle, have the highest

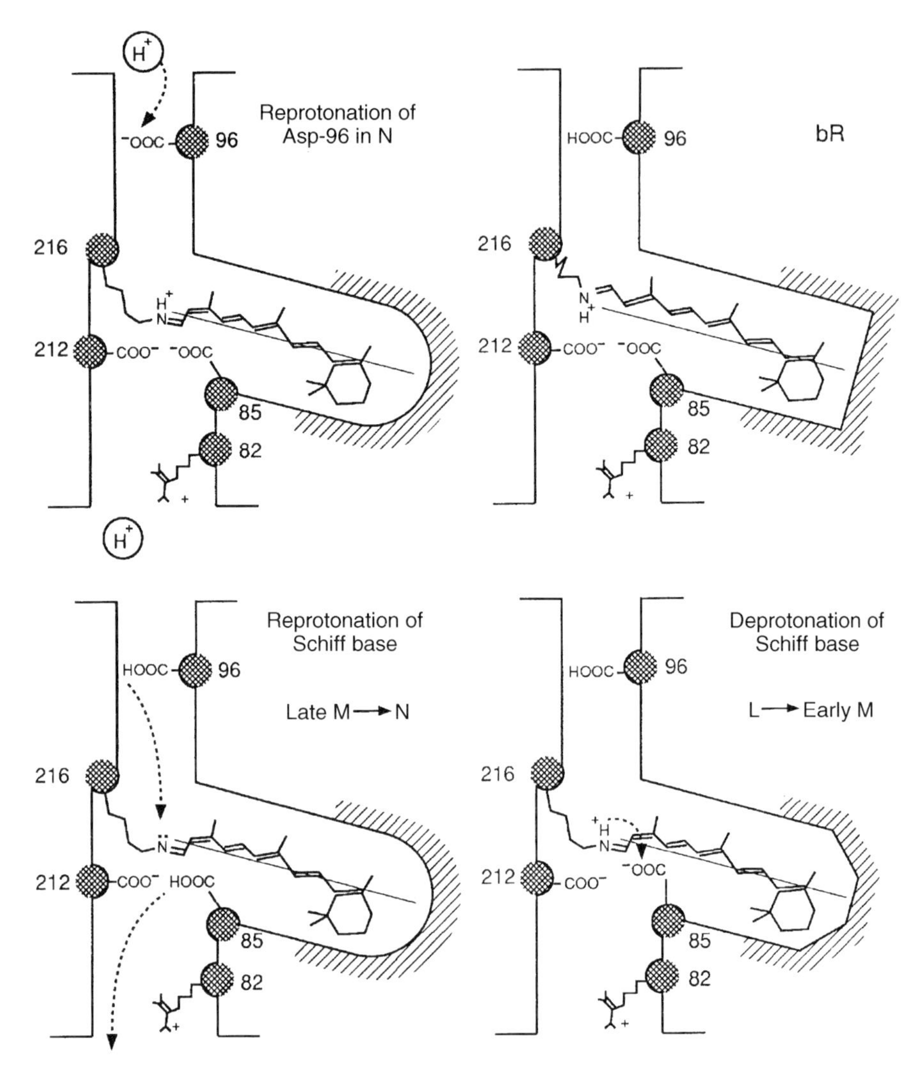

FIG. 3. Schematic mechanism of bacteriorhodopsin showing a sequence of conformational changes and proton movements which underlie proton pumping.

affinity. However, although they get a reasonable agreement, everything is delicately dependent on the assumptions, so that small changes within the error of the model give answers that vary these affinities over quite a large range. Thus the answers are not proper explanations: they are attempts to show we are working in the right ball park.

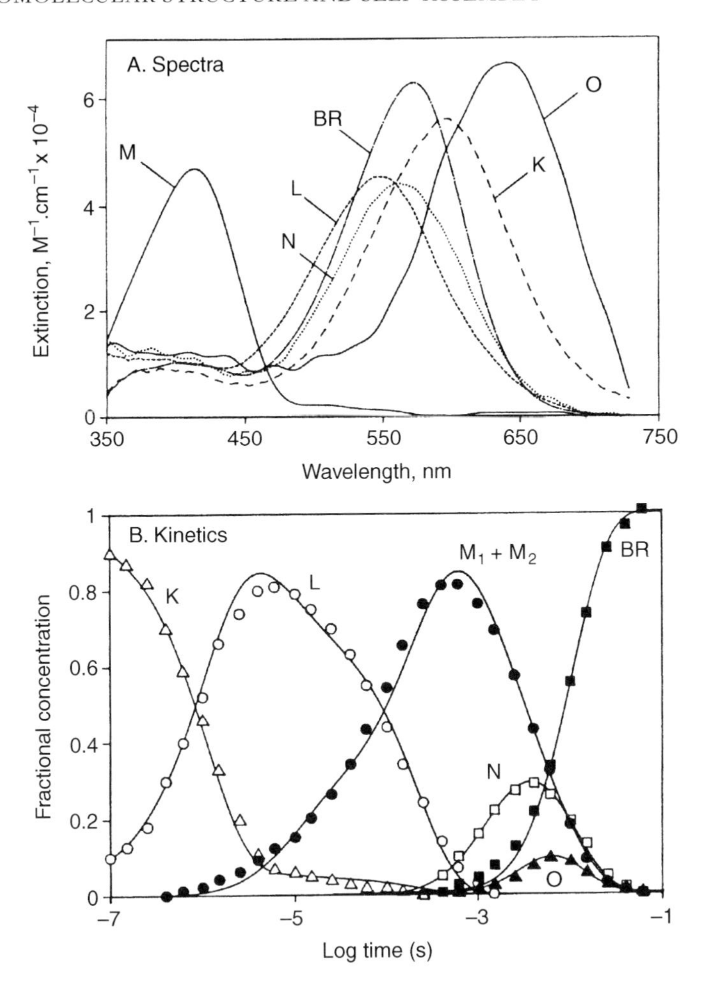

FIG. 4. Logarithmic time plot showing the rise and fall of the sequence of intermediates K, L, M, N and O which form in a linear time sequence during the bacteriorhodopsin photocycle (from Lanyi 1992).

F_1-ATP synthetase

So far we have explored the different levels of explanation in the function of a single protein molecule. Now I would like to move on to look at the membrane F_1-ATP synthetase, which consists of a stalk and a head made up of 10–12 different proteins (Fig. 6). The stalk and head are called F_1 whereas the part in the membrane is called F_0. ATP is either hydrolysed in the cytoplasm and results in protons being pumped

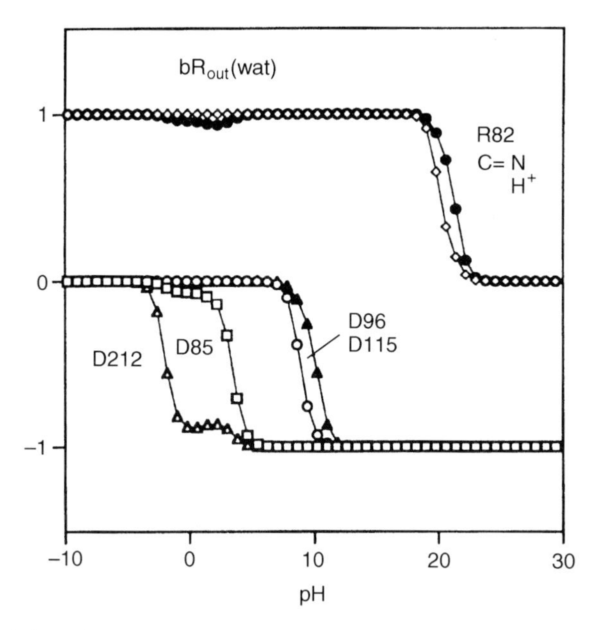

FIG. 5. Simulation of protonation behaviour of a computer model of bacteriorhodopsin as the different carboxylate groups in the structure are titrated (from Sampogna & Honig 1994).

outwards through the membrane, or protons flowing back through the membrane synthesize ATP. Thus, this is an extremely efficient reversible machine.

The atomic structure of part of this complex has been worked out by John Walker, Andrew Leslie, Jan-Pieter Abrahams and their colleagues at Cambridge (Abrahams et al 1993, 1994) (Fig. 7). It is a structure of about 300 kDa. There are three α subunits, three β subunits and a γ subunit that acts as a shaft through the middle of the structure. They found what to me is an extremely satisfying explanation at an intermediate level. Looking down from the top, which is looking down from the cytoplasm to the membrane, they found that the γ subunit is in the middle and the α and β subunits go round in pairs with ATP in the β subunit of one pair, ADP in the next and then an empty subunit. The conformations of all the protein side chains around each of the binding sites is known. The structures are not identical, resulting in an asymmetric structure with three different physical states for the chemically identical subunits. When ATP is hydrolysed and protons are pumped, it comes off and is replaced by another ATP. In the middle there is a single γ subunit surrounded by three chemically equivalent but structurally different subunits. The hypothesis Leslie, Abrahams and Walker made, just from looking at the structure, was that the γ subunit rotates as the ATP is hydrolysed. A Japanese group (Noji et al 1997) tested

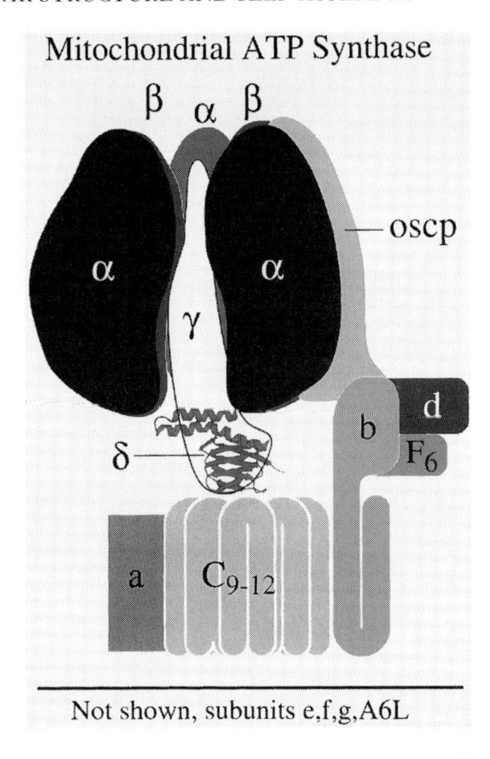

FIG. 6. Schematic diagram of ATP synthetase, showing most of the different protein subunits (courtesy of John Walker).

this hypothesis in an *in vitro* experiment where they took the $\alpha_3\beta_3\gamma$ construct, attached it via the $\alpha_3\beta_3$ end to a glass side using a polyhistine tail, biotinylated the γ subunit on the other end, attached a fluorescein-labelled biotinylated actin filament using streptavidin, then added ATP and looked at what happened to the actin filaments using fluorescence microscopy (Fig. 8).

They found the hypothesis was true and the subunit does rotate. The direction of the rotation predicted simply by looking at the structure is exactly what is found experimentally (Fig. 9). So here there is a connection between the atomic structure and the macroscopic behaviour you can see with your eyes via a very simple hypothesis. In the membrane this γ subunit then makes contact with other protein subunits, and the addition of ATP to the F_1 assembly drives the rotation of the γ subunit which then reaches down into the membrane and by an unknown mechanism is coupled to the pumping of the protons. Therefore this ATPase is acting in a similar way to a motor with a driveshaft and a generator at the other end. This is an adequate explanation at a certain level without us having to go any further.

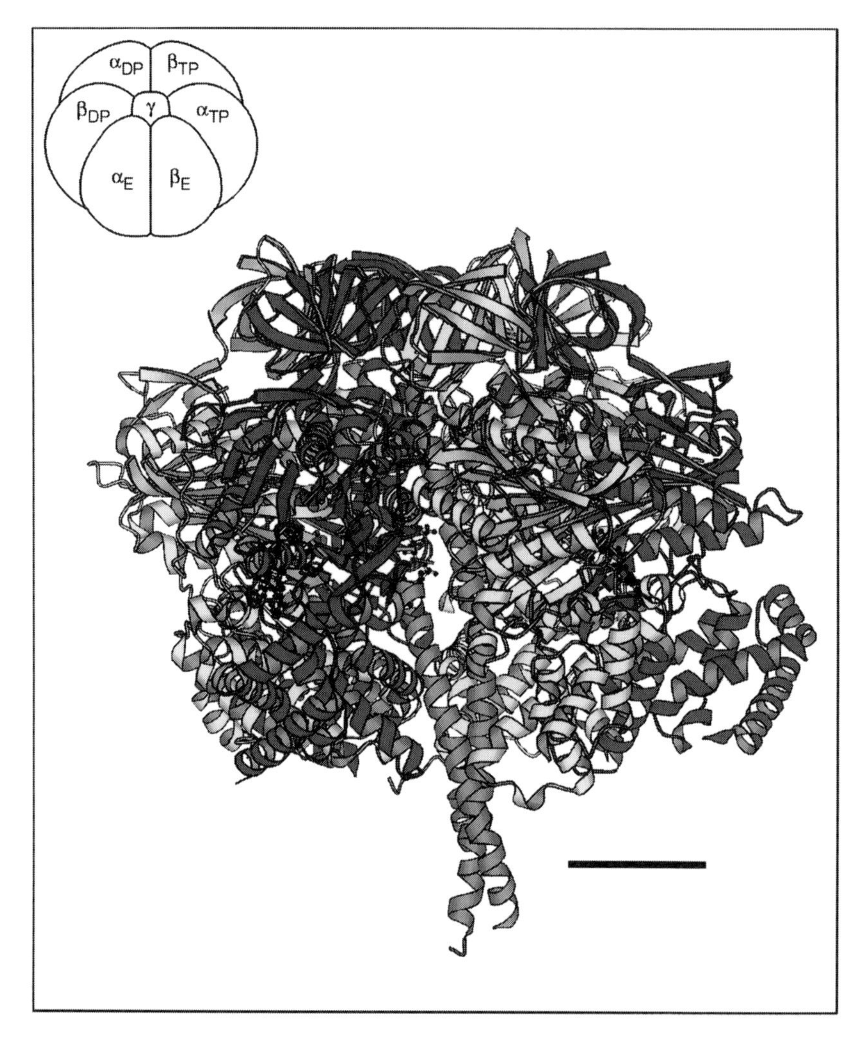

FIG. 7. Atomic model of F_1-ATPase obtained by X-ray diffraction (from Abrahams et al 1994). Scale bar = 20 Å.

Flagellar rotary motor

The flagellar rotary motor that drives the bacterium physically through the medium, is an even more complex structure. It is coupled to cellular sensing, for example, through phototaxis via SRI or SRII or through chemotaxis in response to a variety of chemicals.

Figure 10 shows a picture of a bacterium with several flagella each connected to a structure, the rotary flagellar motor, which is shown schematically in Fig. 11 (taken

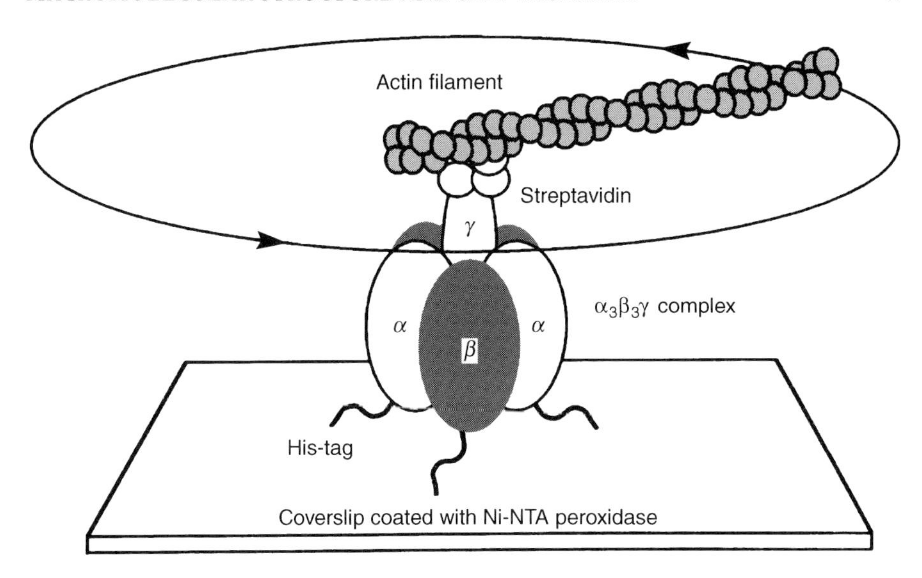

FIG. 8. Schematic diagram showing the underlying idea behind the experiment of Noji et al (1997).

from Namba & Vonerviszt 1997). The isolated complex can be examined by electron microscopy after rapid freezing in amorphous ice. After averaging several independent pictures, it is possible to produce a picture of the average cross-section through the centre of such assemblies and this is shown in Fig. 12 (DeRosier 1995). The structure is an intricate assembly of many different proteins and the exact mechanism by which the energy from the proton gradient is converted into rotational energy to drive the bacterial motor is not yet known. However, it is likely that the mechanism will be similar to the mechanism by which the proton inflow through the F_0 part of ATP-synthetase drives rotation of the γ subunit. In the case of the rotary flagellar motor, the central shaft will rotate, in either direction but controlled by the state of a control protein, clicking through a number of positions defined by the symmetry of the static part of the motor anchored in the membrane.

The mechanism by which bacteria move can therefore be explained at a number of levels. At one level, the existence of a rotary flagellar motor driving the rotation of the flagellum explains how it is possible for the bacterium to move. At the next level, the structure and composition of the motor, made up of more than 40 different protein subunits with a description of which ones move and which ones are stationary, begins to describe the nature of the motor itself. At a third level, not yet reached, the components of the motor directly responsible for the energy

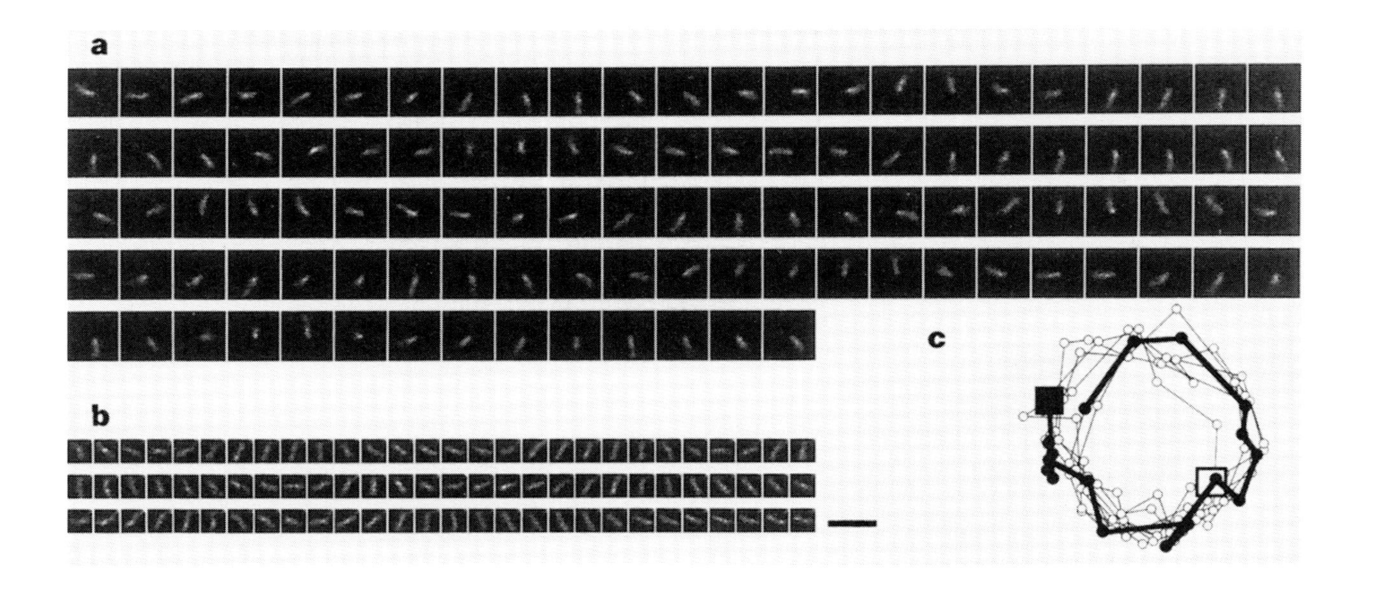

FIG. 9. Time frames showing fluorescence microscopy of the rotating labelled F_1-ATPase γ subunit (from Noji et al 1997).

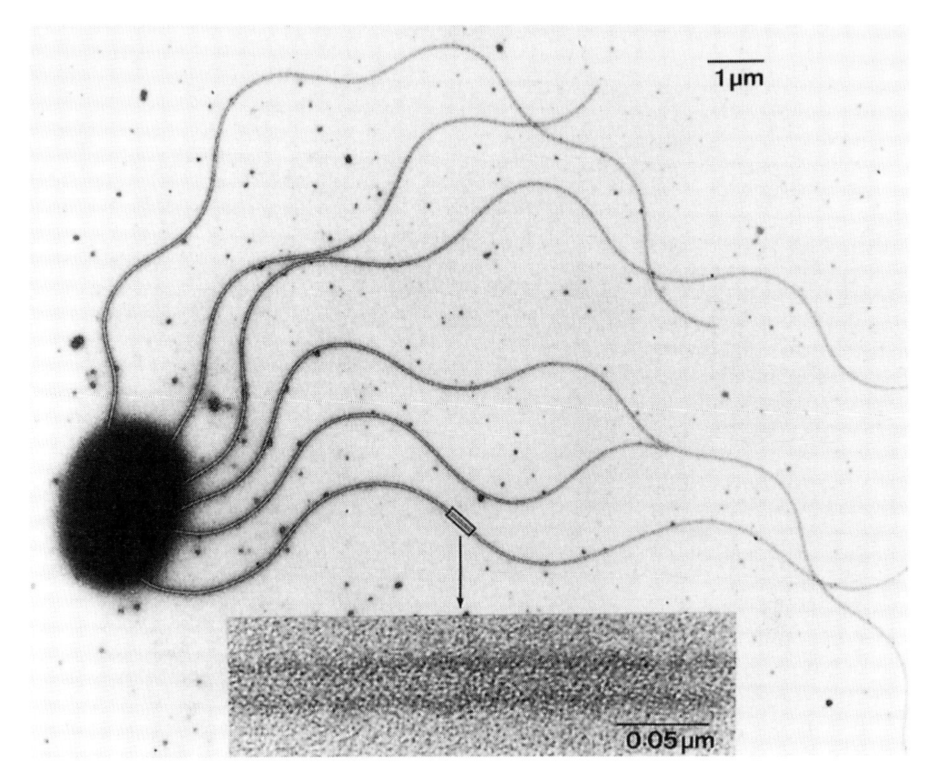

FIG. 10. Micrograph showing flagella in *Salmonella typhimurium* (courtesy of Keichi Namba, from Namba & Vonerviszt 1997).

transduction will be identified. Finally — and at this level my own curiosity would be completely satisfied — it should be possible to explain quantitatively, in terms of the energy input and output with a plausible mechanism for its conversion, how the motor works in detail.

Conclusions

So, what are the limits of reductionism in macromolecular structure? It seems to me that if we carry on working we can produce an explanation at any required level of detail, limited only by the error level of our measurements. Different people working in this area will find the different levels of explanation more or less attractive to them. Fortunately, we have new generations of chemistry graduate students that can take over from previous generations of biochemistry graduate students, and beyond them we have physicists and computer scientists. The only

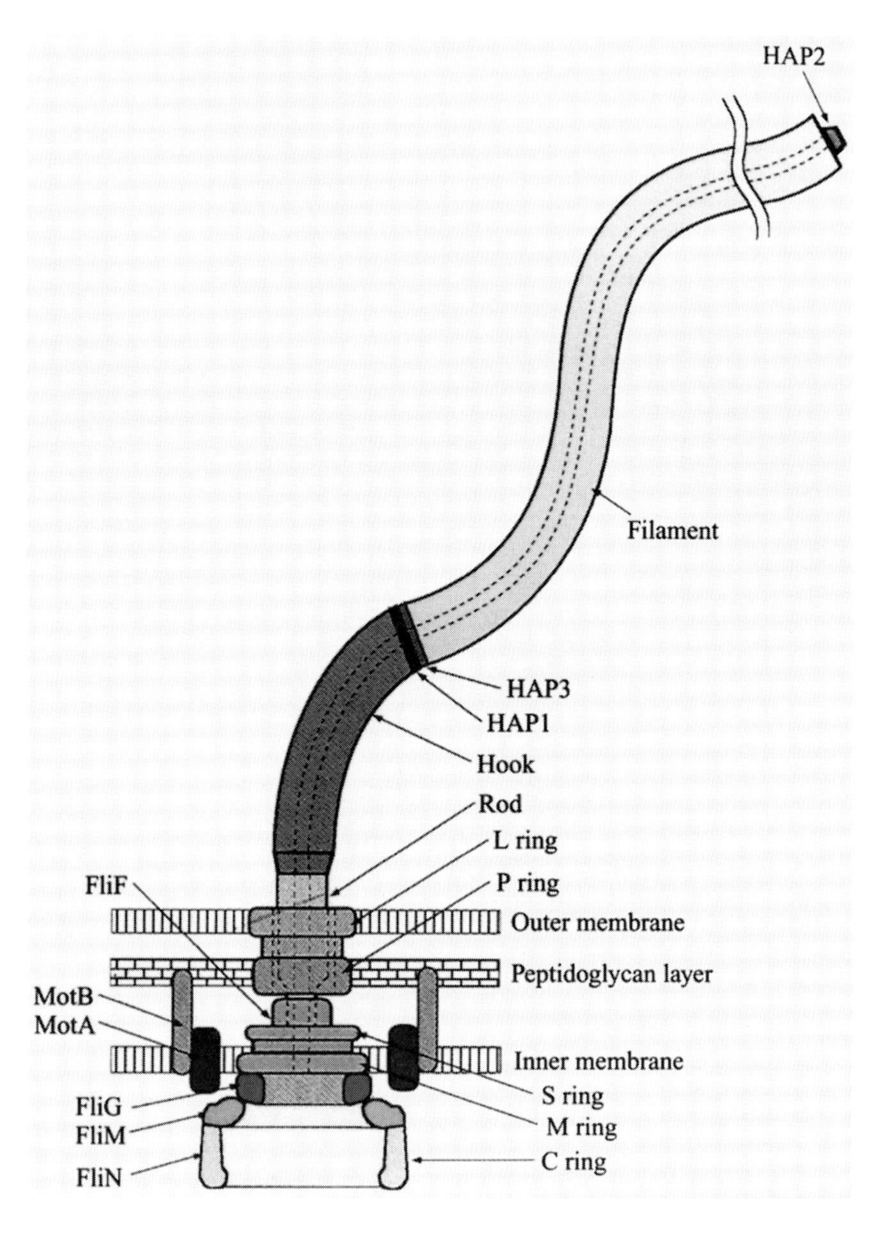

FIG. 11. Schematic of flagellar rotary motor (courtesy of Keichi Namba, from Namba & Vonerviszt 1997).

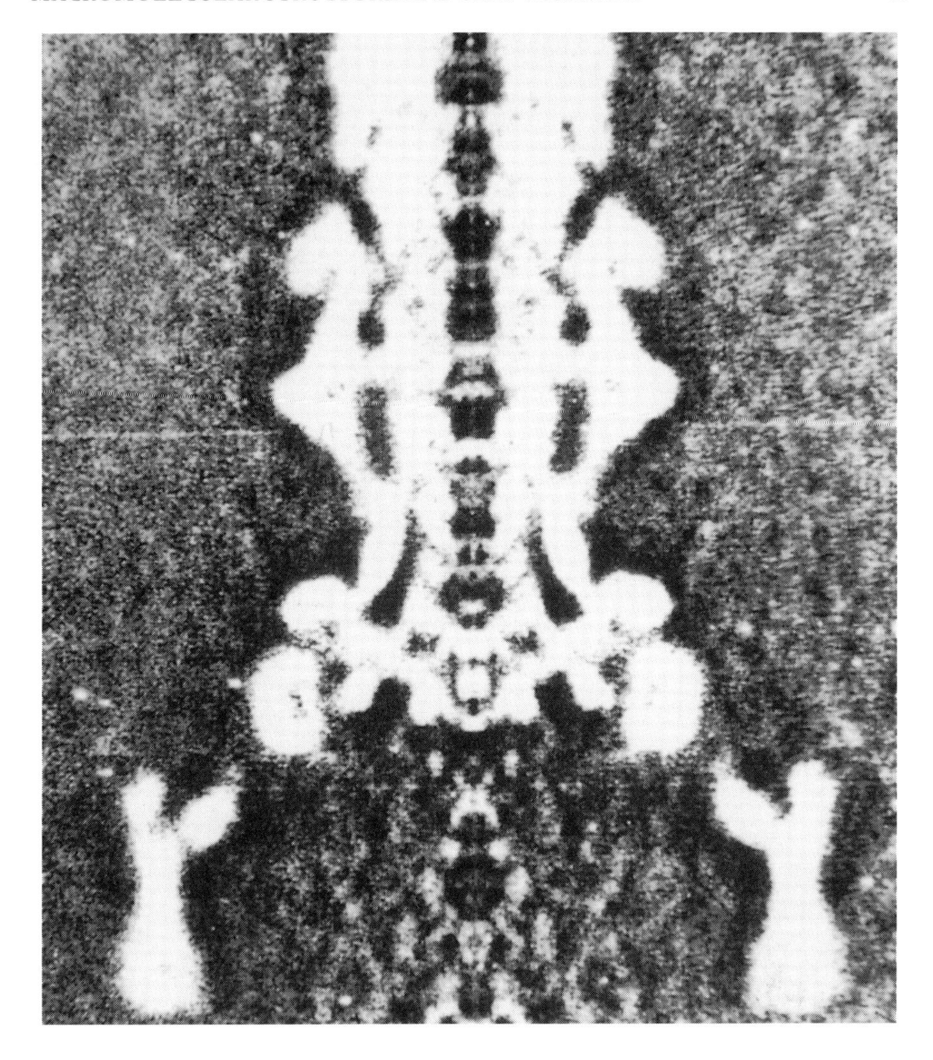

FIG. 12. Electron cryo-microscopic picture of the cross-section through the centre of the average motor assembly (from DeRosier 1995).

thing which limits the interest of a particular group is when their level of interest drops below their boredom threshold, but then another group of people may be available, approaching the problem from a different perspective, who might get quite excited about the topic. In this sense, I suppose I am a reductionist, and I believe that you can explain phenomena at any level you want — you simply aim to fit it in with your own values and become satisfied when your own level of explanation is reached. Boredom therefore sets an individual limit to the appeal of reductionism as a goal in biology.

References

Abrahams JP, Lutter R, Todd RJ, Van Raaij MJ, Leslie AGW, Walker JE 1993 Inherent asymmetry of the structure of F_1-ATPase from bovine heart mitochondria at 6.5 Ångstrom resolution. EMBO J 12:1775–1780

Abrahams JP, Leslie AGW, Lutter R, Walker JE 1994 Structure at 2.8 Ångstrom resolution of F_1-ATPase from bovine heart mitochondria. Nature 370:621–628

Blanck A, Oesterhelt D 1987 The halo-opsin gene 2. Sequence, primary structure of halorhodopsin and comparison with bacteriorhodopsin. EMBO J 6:265–273

Blanck A, Oesterhelt D, Ferrando E, Schegk ES, Lottspeich F 1989 Primary structure of sensory rhodopsin-I, a prokaryotic photoreceptor. EMBO J 8:3963–3971

Blaurock A, Stoeckenius W 1971 Structure of purple membrane. Nature New Biol 233:152–155

DeRosier DJ 1995 Spinning tails. Curr Op Struct Biol 5:187–193

Grigorieff N, Ceska TA, Downing KH, Baldwin JM, Henderson R 1996 Electron-crystallographic refinement of the structure of bacteriorhodopsin. J Mol Biol 259:393–421

Henderson R, Baldwin JM, Ceska TA, Zemlin F, Beckmann E, Downing KH 1990 Model for the structure of bacteriorhodopsin based on high-resolution electron cryomicroscopy. J Mol Biol 213:899–929

Lanyi JK 1992 Proton-transfer and energy coupling in the bacteriorhodopsin photocycle. J Bioenerg Biomembr 24:169–179

Namba K, Vonerviszt F 1997 Molecular architecture of bacterial flagellum. Q Rev Biophys 30:1–65

Noji H, Yasuda R, Yoshida M, Kinosita K 1997 Direct observation of the rotation of F_1-ATPase. Nature 386:299–302

Oesterhelt D, Stoeckenius W 1971 Rhodopsin-like protein from purple membrane of *Halobacterium halobium*. Nature New Biol 233:149–152

Sampogna RV, Honig B 1994 Environmental effects on the protonation states of active-site residues in bacteriorhodopsin. Biophys J 66:1341–1352

Schobert B, Lanyi JK 1982 Halorhodopsin is a light-driven chloride pump. J Biol Chem 257:306–313

Seidel R, Scharf B, Gautel M, Kleine K, Oesterhelt D, Engelhard M 1995 The primary structure of sensory rhodopsin-II: a member of an additional retinal protein subgroup is coexpressed with its transducer, the halobacterial transducer of rhodopsin-II. Proc Natl Acad Sci USA 92:3036–3040

DISCUSSION

Rose: There are two questions I would like to probe you on a little further. The first concerns degrees of freedom: there is an awful lot of protein structure there, a considerable proportion of which is presumably not absolutely relevant to function. This produces a limit to predicting function from structure. The other point is to put the question the other way round, because in an evolutionary context, I would suspect that what constrains structure is function. The issue then becomes what can one predict about structure in defining a function, and that's a different sort of question. This is an antireductionist question because it is arguing that the whole (the function of the organism) constrains the part (that particular protein).

Henderson: In proteins, the exact structure probably doesn't matter. We know from experimental tests of evolution and making purposeful mutations, that you can make many changes without affecting the function of a protein. Also, proteins can be made that are far more stable than those we actually have: this is because in the organism they have been designed not for one purpose but for many—not just to have the catalytic activity, but also to be able to be degraded and built at the right rate. This comes back to the question of how robust the system is. I actually think it is very robust, so you can make great changes in the details of one particular protein and still retain the function. If you then transplant it into a different physical or biochemical environment it will probably perform its function quite well even though it may have moved a long way. The entire system is very robust at every level.

Garcia-Bellido: How much noise or tolerance does the system have? If it has a lot, then the constraints are in the structure, not the function. Or, to put it another way, the function accommodates the possibilities of the construction process. To say that the system is robust and you can do anything you like with it is perhaps a little naïve, because the constraints of the generation process are tremendous.

Henderson: I disagree. It is true that there is noise, but as long as you have a little bit of signal, if that signal has some advantage you can gradually grow out of that a less noisy system under the selection pressure of evolution. You can superimpose quite a number of rounds of additional selection if noise is a problem.

Holmes: Structure is much more highly conserved than function in protein molecules. The fact that there exist only a limited number of protein folds suggests that it is very difficult to design a novel protein molecule. Consequently, these folds frequently change their function. A tremendous constraint in evolutionary biology is speciation of molecules: how do you create a new function with your existing band of molecules? Nobody knows how to get chymotrypsin starting from haemoglobin, for instance. We don't know which pathways led to what we now see, but we know that the number of such pathways is limited.

Garcia-Bellido: There are only in the order of 1000 types of protein folds. Thus the system is constructed by multiplication, iteration and combination. The generative constraints are fundamental.

Hess: We must not forget that there are systems in biology which extract signals by stochastic resonance out of the noise.

With regard to bacteriorhodopsin and photosynthesis, it is interesting that all the biological systems that interact with light are designed to trap energy with an amazing efficiency. This is obviously the result of an evolutionary fitting procedure, but it is maintained from the lowest to the highest organisms. It would be very interesting to study and understand the history of this type of evolution. No system in the artificial chemical world can achieve this type of energy-trapping efficiency.

Henderson: I don't agree that bacteriorhodopsin is particularly efficient; it could be a lot more efficient. After all, in evolution it has been replaced by photosynthesis in green plants using chlorophyll-based absorption of light. This would apply to any other protein, in spite of what people such as Jeremy Knowles have said about certain enzymes, such as the glycolytic enzyme, triose phosphate isomerase, being perfect. They are perfect by one definition, but they have to respond to a variety of pressures in life, including being able to be synthesized and degraded. Taking everything into account, we may at this moment be in a sort of metastable state from which it's hard to get out of, but that doesn't mean that you can't make a more efficient protein.

Perutz: With regard to Richard Henderson's point about how robust the systems are, the haemoglobins are widely distributed in vertebrates and invertebrates. The globin fold is achieved with about 140–150 amino acids. Of these, there are only two that have never been replaced: the histidine that is linked to the iron, and the phenylalanine that holds the haem in place. All the others have been substituted by different amino acid residues in different species, and yet the three-dimensional fold is remarkably similar. There is extraordinary adaptability. In addition, the oxygen affinities of different haemoglobins differ by a factor of up to 100 000, so the protein again can adapt to meet the particular needs of the organism.

Raff: I want to return to the question Richard raised about what is satisfying or interesting, and whether it really is just a question of levels of analysis. It seems to me that it is not the level of analysis that's important, but the novelty of the findings. I suspect that we can all be interested in any level of analysis if the answer is novel and important enough.

Noble: The difficulty with this is that what is exciting to one person is not to another, and what is in need of an explanation to one is not to another. The dilemma this creates for us is that the whole issue becomes to some extent an educational and political issue. This is not a question that in itself can be resolved scientifically.

Dover: Let me address the question of what scientists find interesting. Obviously money is a major factor here, but I think the do-ability of the project also plays a large role in the choice of what we study. For example, your studies on *Halobacterium* worked because what you're handling are simple and direct relationships between a molecule and the phenotype, between the structure and the function. Once you get into more complex systems these relationships quickly collapse. Let me give an example of this. People have tried to model, mathematically, how a single-celled *Drosophila* egg can change into a multicellular system, by considering the well known gradient of the Bicoid protein. This gradient runs from the top of the egg to the bottom and, at a given concentration, turns on a gene called *hunchback*, which is turned on in a stripe. The relevant question here concerns how a gradient of gene expression can be turned into a stripe of the same expression. This can be described mathematically in terms of the number of binding sites in front of *hunchback* that Bicoid binds to, their binding affinities, the cooperativity between Bicoid molecules and so on. This can be modelled to a reasonable degree of precision and looks fine until you try to go a little bit further, and then you realise that both Bicoid and Hunchback are involved in multiple other functions and that pleiotropy (one gene involved in many other functions) and epistasis (many other proteins involved in the same function) are involved to such a degree that it is not possible to model this system. Because there is such a high degree of pleiotropy and epistasis in complex biological systems they cannot be modelled in a reductionist manner unless you restrict yourself to looking at one little bit at a time. This doesn't mean, however, that we have to go to the other extreme of saying that it is the whole organism and its relation to the environment that will ultimately decide how, where and why Bicoid evolved to interact with Hunchback.

Henderson: In my paper I tried to keep what I was saying within the disciplinary boundaries between chemistry and the cell, but what you're saying is broadening the issue out again. It comes back to an insight from Thomas Nagel's paper (Nagel 1998, this volume), where he said that after many reductionist considerations, maybe natural selection—which closes the circle of function back to structure—bridges the gap between the chemistry and the physics and what you end up with in your organism. Thus even if in the middle you change the behaviour of some of the components such as at your Bicoid or Hunchback stage, and you then put on your natural selection at the end, the whole thing is so robust that you could tolerate all sorts of changes, provided you have closed the cycle in terms of a self-replicating organism that could reproduce, mutate and be selected for.

Reference

Nagel T 1998 Reductionism and antireductionism. In: The limits of reductionism in biology. Wiley, Chichester (Novartis Found Symp 213) p 3–14

Reduction and integration in understanding the heart

Denis Noble

Department of Physiology, University of Oxford, South Parks Road, Oxford OX1 3PT, UK

Abstract. The heart provides an excellent example of the limits of the reductive approach. Cardiac cells function through the interaction of a very large number of ion transporters, and the processes that link these to metabolic states and to contraction. Yet, the great majority of the advances made recently have been at the cellular and molecular levels. The pressing problem now is to begin to understand the highly complex interactions that create physiological function at a cellular level and, in turn, to understand the way in which large numbers of cells interact to produce the activity of the whole heart. Many kinds of arrhythmia, for example, can only be understood at the whole organ level. Successful interventions using drugs designed to treat cardiac disease depend on an integrative understanding, which at present we do not have. This is one of the reasons why clinical trials of drugs treating cardiac arrhythmias have been spectacularly disappointing. This paper illustrates some of these problems by analysing normal and abnormal heart rhythms, and by focusing on one particular transporter, the sodium–calcium exchanger, that is deeply involved both in normal calcium balance in the heart and in the generation of pathological states, including life-threatening arrhythmias. It will be shown that some surprising counterintuitive results appear when computations are done at an integrative level.

1998 The limits of reductionism in biology. Wiley, Chichester (Novartis Foundation Symposium 213) p 56–72

The limits of the reductionist approach in biology are only too evident to us today precisely because the reductionist programme has been so immensely productive and has therefore in turn created both a major challenge to integrative work and an immense opportunity. The unravelling of genetic code, of molecular structure, of subcellular mechanisms, has been so breathtakingly rapid in creating a mountain of detailed information that integrative work has barely had time to define the problems let alone tackle them on the scale required. Nevertheless, the integrationist agenda is being defined (see for example Boyd & Noble 1993). I will return to these general issues at the end of this paper.

The heart provides an excellent example both of the huge advances that have been made using the reductive approach and of the severe limits of this approach when attempting to understand the complex phenomena of normal and abnormal rhythms.

The natural cardiac pacemaker

Consider first the mechanism of normal rhythm. Natural heart rhythm is generated in a small specialized region called the sinoatrial (SA) node. The cells in this region generate spontaneous depolarizations, called the pacemaker potential, so that the membrane potential automatically reaches the threshold for initiation of action potentials. There is, clearly, an oscillator at work here. But at what level does it operate? Atomic? Molecular? Subcellular? Cellular? Multicellular? Even at the level of the whole organ and its nervous innervation? All are clearly possible in principle, though there are obvious reasons for thinking that the first two would have difficulty with the relatively low frequency of cardiac rhythm. All the others have, though, been seriously proposed at one time or another.

An example of multicellular rhythm would be that of re-entry as propagated action potentials chase around a closed loop following a pathway of interconnected cells. Many forms of arrhythmia are of this nature, including the life-threatening arrhythmias of tachycardia and fibrillation which occur during a heart attack (Wit & Janse 1993). It is easy to show that normal rhythm is not of this nature. When we isolate individual sinus node cells, they each individually show pacemaker activity. The basic oscillator must therefore exist at the cellular level or below.

We can also show that it does not exist below the cellular level. This is surprisingly easy to do. If we control the cell voltage by applying a feedback system known as the voltage clamp, we find that no matter at what level we hold the membrane potential, the ionic current mechanisms do not oscillate. Each will display monophasic or biphasic responses to voltage changes as the channels and transporters responsible open and close, but all then move rapidly to a steady state. This tells us not only that each of these mechanisms individually does not oscillate but also that, under normal circumstances, no subcellular mechanism oscillates that could generate surface membrane current changes.

This tells us that a primary analysis of natural cardiac rhythm can be done by first recording the opening and closing kinetics of the individual ion transporters, in particular sodium, calcium and potassium. If we then set up systems of differential equations accurately describing these kinetics and then incorporate them all into an integrated description of the cell, we should be able fully to reconstruct the pacemaker rhythm. It should emerge as a global property of the complete system of ionic transporters. This is the case, as shown in Fig. 1. We can use such models to investigate further integrative properties, such as the reciprocal role that some of the components of current play in stabilizing cardiac rhythm (Noble et al 1992).

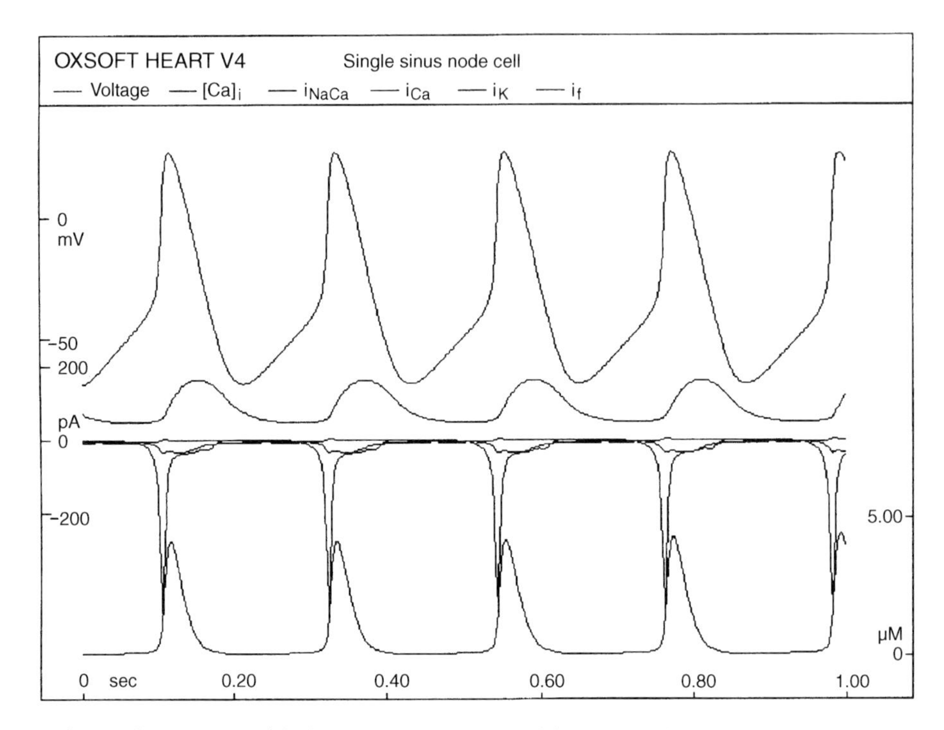

FIG. 1. Computer model of pacemaker activity in rabbit sinus node cell. The top trace shows the membrane potential, which oscillates at a frequency of around 4 Hz, so generating the natural cardiac rhythm. The middle traces show variations in the current carried by a few of the ionic transporters involved: calcium channels, potassium channels, sodium–calcium exchange and the hyper-polarizing activated current. The bottom trace shows the variations in intracellular calcium which are responsible for activating the contractile proteins to produce a heartbeat during each voltage oscillation (Noble et al 1992).

From this you might conclude that the answer to our original question is a very simple one indeed: we carry out a reductive analysis of the subcellular or molecular mechanisms involved in each ion transporter, and then integrate them into a description of the cellular oscillator. Is this sufficient?

One of the limitations of the reductive approach in this case, even when used in conjunction with a form of re-integration (which is what cell-level computer modelling amounts to), is illustrated by the fact that this is by no means the end of the story. One way of appreciating that this cannot be all there is to understanding natural rhythm is to note that when cells are isolated from the SA node, they all beat at different frequencies, depending primarily on their original location within the node tissue (Kodama & Boyett 1985, Honjo & Boyett 1992). The outer regions beat faster and show much larger voltage oscillations than the central regions. Yet the normal heartbeat is a synchronized event in which a single propagated signal emerges from the SA node during each beat.

It is possible to show, using a multicellular model in which large numbers of SA node cell models are connected up together into a network, that these interconnections (called gap junctions) ensure synchronization. A surprising, but very pleasing, result of these computations is that this synchronization can be achieved even with an extremely low density of the gap junction proteins (Winslow et al 1993a). This result is pleasing because it solves an otherwise puzzling fact, which is that the density of these proteins is indeed exceedingly small in the SA node region.

But, the multicellular computations contain even greater surprises than this. First, if we reconstruct the whole node by arranging the model cells to display the spectrum of natural frequencies seen in the isolated cells from each region, we end up with a spectacular failure to reproduce what happens in a normal heart: the wave of excitation propagates in towards the centre, not out towards the rest of the heart! I can well remember the reactions we had when we first obtained this result. Was there something wrong in the computational methods, or with the cellular or molecular information on which the modelling was based? The answer is neither of these. The answer is that, not only do you need to reconstruct the whole node to account for normal rhythm, you also need to connect it up to the atrium. The electrical interactions between the atrial cells and the sinus node are so large that they are responsible for determining the sequence of excitation within the node and for ensuring that the signal propagates outwards towards the atrium itself. Moreover, the fine details of this interaction and the ability of the electrically weak sinus node cells to excite the electrically strong atrial cells depend on exactly how the node tissue is connected to that of the atrium. There are fine interdigitations of each tissue protruding into the other. It can be shown that this arrangement is ideal for ensuring a high safety factor for transmission out of the node into the atrium (Winslow & Jongsma 1995).

Abnormal cardiac rhythms

Cardiac arrhythmias form one of the commonest group of diseases. Moreover, some are seriously life-threatening. During a heart attack, for example, the ventricular tissue contracts in a disorganized way as different regions of the ventricle are excited by multiple waves of re-entry. This is, clearly, a complex phenomenon which necessarily requires analysis at the level of the whole organ for a full understanding. The triggering factors, however, lie at a lower level as small regions of tissue become ischaemic as a consequence of the obstruction of their arterial supply. This deprives the tissue of a re-supply of essential metabolites.

For the last year or two, my laboratory has been collaborating with biochemists, physiologists and computer scientists in a project to reconstruct ischaemic heart arrhythmias. We can attempt this because models of the electrical activity of cardiac cells are now highly developed (Noble 1995). Starting with the DiFrancesco–Noble model of the Purkinje conducting tissue of the ventricles, the latest generation of models include ion concentration changes, the activity of ion pumps and exchangers as well as of ion channels, the buffering, sequestration and release of internal calcium, and the activation of contraction (DiFrancesco & Noble 1985). These developments have already greatly extended the range of application of the models, in particular to an understanding of the cellular basis of many forms of arrhythmia (Noble 1991). The cell models have also been incorporated into massive network simulations of cardiac tissue (Winslow et al 1993b). With the rapid increase in computing power, it has become clear that simulations of the whole heart using biophysically detailed models is feasible.

This is a large and ambitious undertaking, requiring a particular combination of experimental and computing methods. In addition to the mechanisms already incorporated into the models, changes in the relevant energy metabolites and ion concentrations must be incorporated. Here, I will first briefly summarize two major aspects of the work: the modelling of pH changes and the modelling of metabolite changes.

Modelling of pH changes

Adequate control of intracellular pH in the heart is a fundamental requirement, and deviations from resting levels of around 7.2 result in major changes in contraction and excitability. Under acidic conditions, pH_i in heart is controlled by at least four sarcolemmal ion carriers: two acid extruders (Na^+/H^+ exchange and Na^+-HCO_3^- symport) and two acid loaders (Cl^-/HCO_3^- exchange and the recently proposed Cl^-/OH^- exchange). Each of these transporters has been modelled using detailed experimental data on their pH dependence.

These equations for proton fluxes were then incorporated into the cellular models in conjunction with modelling work on contraction. Figure 2 shows that we can model an acid load of increasing magnitude and observe the effects on contraction and sodium. The changes observed in pH_i and $[Na]_i$, and the effects on contraction are qualitatively correct. Experiments on single cells have shown the negative inotropic effect of acidosis (Bountra & Vaughan-Jones 1989), and this is also seen in the model. The recovery in pH_i after acid load, the rise in internal sodium, and the effects on contraction all reflect experimental findings. Notice also that when pH is reduced to 6.5 a qualitative change in response is predicted: internal calcium oscillations lead to irregular ectopic beating.

A. Simulations with the constant for Ca binding to its release site set at 1 μM

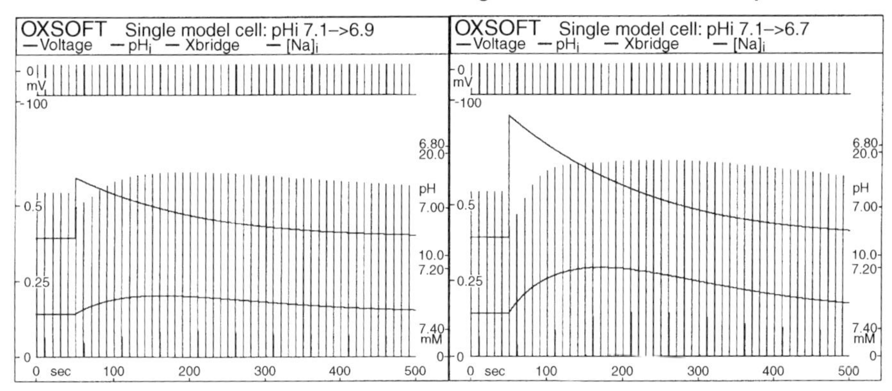

B. Simulations with the constant for Ca binding to its release site set at 4 μM

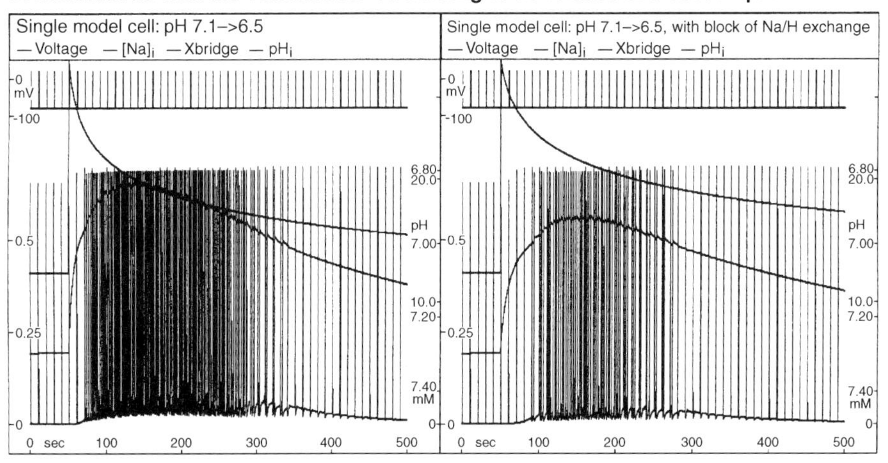

FIG. 2. OXSOFT HEART model: computation of the changes in contraction and internal sodium on acid loading in a model cell. Top trace in each panel shows voltage clamping, stimulating at 10 Hz, while the middle trace represents internal pH; the acid load increases successively from the resting pH_i of 7.1 to 6.9, 6.7 (top pair, A) and 6.5 (B). The trace that rises suddenly with the acid load is that of internal sodium; the greater the pH_i load, the greater the rise. Finally, the bottom trace represents contraction; note the inotropic effect of acidosis in A, and the production of ectopic beats as pH_i is changed to 6.5 in B. The lower pair of computations in B shows the effect of blocking the Na^+/H^+ exchanger, calculated with a higher value for the constant for calcium binding to its release site (Km_{Ca}). The model predicts an attenuation of the arrhythmic beats when Na^+/H^+ is blocked (Ch'en et al 1997).

These computations show that intracellular acidity to a degree that occurs during ischaemia would be expected to generate sodium-overload arrhythmias. What is interesting is that drugs which block Na^+/H^+ exchange should attenuate these, as can be seen in the pair of simulations in Fig. 2B when the flux contribution from the Na^+/H^+ exchanger is set to zero. It is noteworthy that Hoe 694, a blocker of the Na^+/H^+ exchanger, has been shown to be cardioprotective and anti-arrhythmic (Scholz et al 1993).

Are the oscillations generated in this way comparable to the normal pacemaker rhythm? The answer is no. They occur at a different level. If sodium- and calcium-overloaded cells are voltage clamped, we find that the oscillations of calcium continue even though the oscillations of potential are prevented. This tells us that the mechanism is subcellular. It is in fact generated by the interaction between cytosolic calcium and the release of calcium from the sarcoplasmic reticulum (SR). During normal excitation it is the rise in calcium consequent on the opening of calcium channels during the action potential that triggers further, and much larger, release of calcium from the SR. This is an amplifier, and like most amplifiers it is liable to positive feedback oscillation if the driving factor, in this case cytosolic calcium, is increased beyond its normal range.

Modelling of energetics

The model of metabolic changes used three basic equations as its basis:

$$\mathrm{ATP \rightarrow ADP + P_i}$$
$$\mathrm{PCr + ADP \leftrightarrow ATP + Cr}$$
$$\mathrm{2ADP \leftrightarrow ATP + AMP}$$

If the starting concentrations for all the metabolites are known, then it is possible to create equations for total adenine (1), total creatine (2), total phosphate (3), the creatine kinase reaction (4), and the myokinase reaction (5) as follows:

$$[\mathrm{ATP}]+[\mathrm{ADP}]+[\mathrm{AMP}]=X\ \mathrm{mM} \quad (1)$$
$$[\mathrm{PCr}]+[\mathrm{Cr}]=Y\ \mathrm{mM} \quad (2)$$
$$[\mathrm{PCr}]+[\mathrm{Pi}]+3[\mathrm{ATP}]+2[\mathrm{ADP}]+[\mathrm{AMP}]=Z\ \mathrm{mM} \quad (3)$$
$$[\mathrm{ATP}][\mathrm{Cr}]=200[\mathrm{ADP}][\mathrm{PCr}] \quad (4)$$
$$[\mathrm{ADP}]^2=[\mathrm{ATP}][\mathrm{AMP}] \quad (5)$$

These five equations can be rearranged to be solved simultaneously, with the changes in the metabolites a function of the internal phosphate concentration. The modelled changes are plotted in Fig. 3. The model shows the depletion of intracellular phosphocreatine and ATP, while inorganic phosphate, ADP and Na^+ concentrations increase. The changes in metabolite levels are qualitatively correct and correlate well with experimental studies (Clark et al 1987), though other processes have yet to be incorporated. These include anaerobic glycolysis, changes in contractile function, degradation of AMP and changes in pH; their effect will be to change both the end point at which the ATP consumption and production are again equal, and also the metabolite concentrations present at this new steady state.

On reperfusion, blood flow is restored to a heart. To simulate this, we returned the concentration of metabolites to their original levels, with the model otherwise in the same state as at the end of the period of ischaemia, i.e. in a state of Na^+ and Ca^{2+} overload. Re-activation of the SR Ca^{2+} pump in these conditions leads to increased loading of the SR and thus accentuates the oscillatory release of Ca^{2+}. This effect emerges from the modelling as a possible cause of lethal reperfusion injury, since in Fig. 4 we see that the model shows the arrhythmias to be more severe, in agreement with physiological findings.

Suppression of ischaemic arrhythmias

One of the purposes of modelling complex integrative processes on this scale is that if the analysis is complete enough, it should be possible to use such models to predict how to interfere with fatal arrhythmias and, hopefully, design a strategy for their suppression.

There are many precedents for this approach. Anti-arrhythmic drugs are usually classified into classes depending on the particular receptor and channels they act on. This approach depends on identifying which ionic channel or exchanger carries the current responsible for a particular form of arrhythmia and then looking for drugs that block the current. In the case of ischaemic heart disease, the record of success is very poor. Recent clinical trials (such as CAST) have thrown very serious doubt on this approach. Many of the drugs thought to be effective have turned out to be worse than placebo. Part of the problem lies in the fact that the classification of anti-arrhythmic drugs pays little attention to the complexity of the heart. There is in fact no guarantee that, even if one has identified the ionic transporter immediately responsible for the current generating the arrhythmia, blocking this current will be effective in countering the arrhythmia. The transporter may play other roles and interact within the complex system of cells and their interconnections in a way that is difficult to predict from a reductionist approach.

This point can be illustrated by considering one of the transporters known to be involved in sodium–calcium overload arrhythmias: the sodium–calcium exchanger.

The immediate cause of the ectopic beats shown in the reconstructions illustrated in Figs 3 and 4 is that each oscillation of intracellular calcium causes the sodium–calcium exchanger to pump calcium out of the cell in exchange for sodium. Since three sodium ions enter for each calcium ion extruded, this mode of operation of the exchanger generates an inward current. This is responsible for depolarizing the membrane to the threshold for initiating excitation and so for generating ectopic

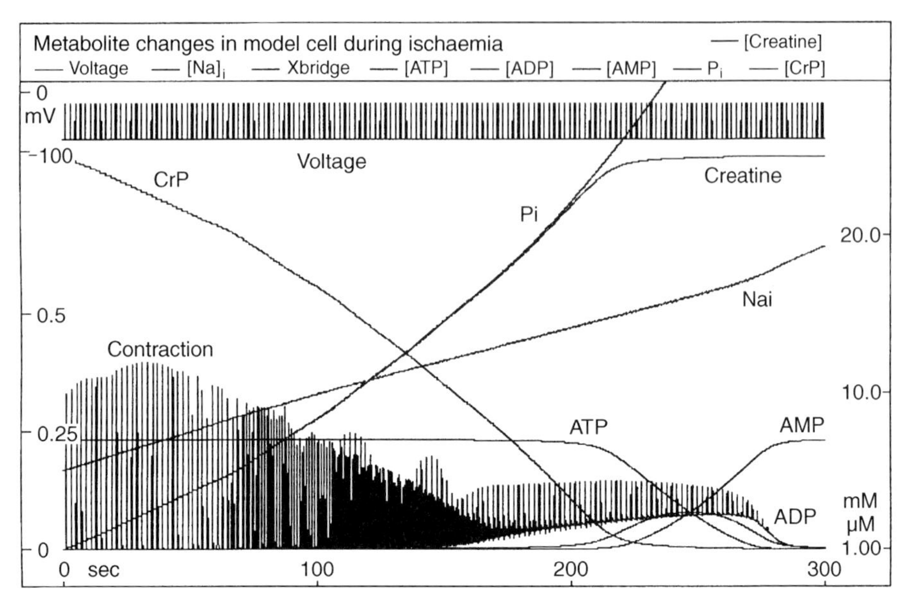

FIG. 3. OXSOFT HEART model: computation of the changes in metabolites during five minutes of ischaemia. Voltage clamping with stimulation of 10 Hz. This modelling succeeds in reconstructing the main changes during ischaemia, including the time course and sequence of metabolite changes, the decline in contraction and eventual contracture, sodium–calcium overload and the occurrence of arrhythmias (starting at around 75 seconds) (Ch'en et al 1997).

beats. If the region involved is large enough (see Winslow et al 1993b) the beat will propagate into the rest of the heart and may trigger a re-entry arrhythmia.

This leads to the prediction that if an inhibitor of sodium–calcium exchange were to be applied during the period of ectopic beating, the arrhythmia would be suppressed. This is the case, and pharmaceutical companies have already developed sodium–calcium exchange inhibitors (Watano et al 1996).

Yet, this clear prediction is spectacularly misleading! If we re-run the computations shown in Figs 2, 3 and 4 with sodium–calcium exchange inhibited, the arrhythmic episodes are even more severe, not less. Paradoxically, to suppress the arrhythmias we need to increase the activity of the sodium–calcium exchanger (Noble et al 1997). This result is illustrated in a simplified model in Fig. 5.

The reason for this surprising result takes us back to the heart of the tension between reductive and integrationist approaches to thinking about a complex phenomenon such as heart arrhythmia. The sodium–calcium exchanger is a paradigm example of a transporter that plays multiple roles in the control of calcium balance. In the heart, its primary physiological function is to pump calcium rapidly out of the heart following each heart beat.

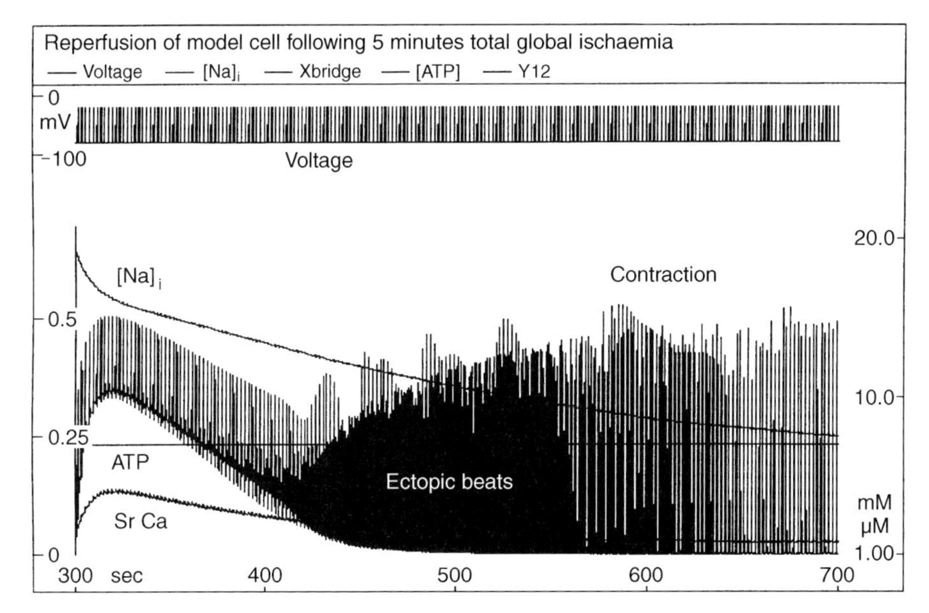

FIG. 4. OXSOFT HEART model: computation of the effects of reperfusion following five minutes of ischaemia under conditions of voltage clamp. This is a continuation of the modelling shown in Fig. 3 but with the metabolites restored and maintained at normal levels to mimic reperfusion. Note that the period of ectopic arrhythmias is more extensive than during the ischaemic period, as observed experimentally (Ch'en et al 1997).

Conclusions

It is by no means accidental that the illustration of the integrative approach that I have used in this paper involves heavy use of computer models. The limits of the reductionist approach are not only that it will fail to account for functional phenomena at a higher level, or that function will often only become evident at a higher level. There is also a technical limit, which is that the network of interactions of individual components in biological systems are frequently so complicated that understanding them requires computing tools, partly in order to explore the interactions quantitatively in a reasonable period of time, and also because, beyond a certain level of complexity, 'armchair' theorizing becomes inadequate to the task. Quantitative, computational biology representing physiological function is therefore set to make a major contribution. This development is central to the way in which physiology is now developing.

Physiology's role as a science must, today, be carved out of its relationships to the genome, to molecular biology and to modern evolutionary biology. That was the theme of the book, *The logic of life*, I referred to in the introduction. Cast in the

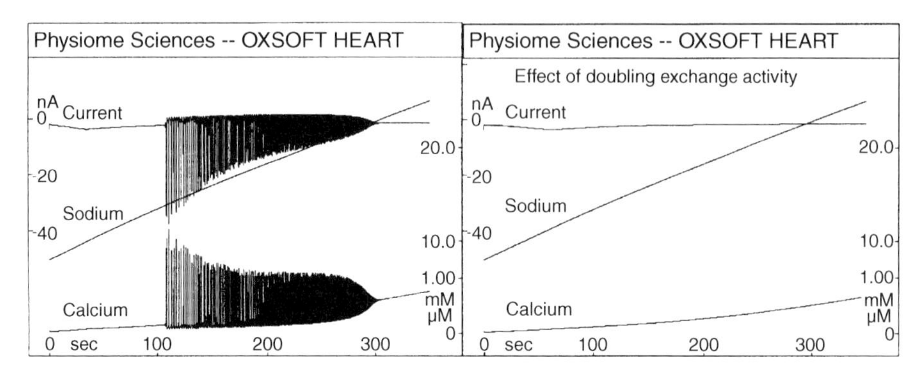

FIG. 5. Modelling of oscillations of intracellular calcium and sodium–calcium exchange current during sodium-overload. Left: this simulation uses the standard OXSOFT atrial cell model (Earm & Noble 1990). Sodium-pump inhibition was used to generate a rise in internal sodium. When the rise in sodium becomes large enough, internal calcium oscillations occur, which generate oscillations in sodium–calcium exchange current. Right: these oscillations are eliminated when the activity of the sodium–calcium exchanger is doubled (Noble et al 1997).

form of a polemic, and echoing Sir James Black's now famous (or infamous, but certainly mischievous) remark that the future lay in the 'progressive triumph of physiology over molecular biology', it turned the selfish gene argument on its head to depict genes more as 'captives' of the successful physiological systems that carry them from one generation to another. This was 'purple prose', of course, chosen deliberately to have immediate impact. But, how else do you restore the balance in a debate that has swung a pendulum far too far in one (reductionist) direction?

Purple prose — choosing colourful language for a greater effect — can indeed be necessary, but it also has its dangers. In particular, it needs to be understood in the context in which it was written. It is not too difficult to imagine a different context in which some of the colourful language of *The logic of life* will appear unnecessary and even misleading in its turn, just as many of the claims made for molecular biology and for the genome project have been.

Physiology, after all, cannot 'triumph' over anything (except perhaps ignorance and superstition), least of all over a valuable tool like molecular biology; it can only interpret it. That itself is, in any case, a greater achievement than conquest. To return to colourful language again: if the genome contains all the 'information' for physiological function to emerge, then physiology contains all the 'interpretation' necessary to understand the genome. Some have even gone so far as to emphasize this symbiotic relationship by calling the quantitative description of physiological function the 'physiome', as a deliberate linguistic echo of the genome project.

Moreover, the interpretation will be assimilative not eliminative, as are all successful absorptions of a new culture by an older one. The assimilated culture becomes an integral part, creating something that is far stronger than either could be alone.

That is why physiology has such a rich and adventurous future. The genome, molecular biology and modern evolutionary biology have all injected new life into physiology. Integration is much more powerful than it could ever have been without these tools and the vast databases of information that they have created. And powerful computing has arrived at just the right time to enable us to exploit this opportunity.

Acknowledgements

This work was supported by the British Heart Foundation, Medical Research Council and the Wellcome Trust. The computer modelling used OXSOFT HEART 4.8 and was supported by Physiome Sciences Inc.

References

Bountra C, Vaughan-Jones RD 1989 Effects of intracellular and extracellular pH on contraction in isolated mammalian cardiac muscle. J Physiol 418:163–187

Boyd CAR, Noble D 1993 The logic of life. Oxford University Press, Oxford

Ch'en F, Clarke K, Vaughan-Jones RD, Noble D 1997 Modelling of metabolites and ion concentration changes during cardiac ischaemia. Adv Exp Med Biol 430:281–290

Clarke K, O'Connor AJ, Willis RJ 1987 Temporal relation between energy metabolism and myocardial function during ischaemia and reperfusion. Am J Physiol 253:H412–H421

Clarke K, Anderson RE, Nedelec JF, Forster DO, Allay A 1994 Intracellular and extracellular spaces and the direct quantification of molar intracellular concentrations of phosphorus metabolites in the isolated rat heart using ^{31}P NMR spectroscopy and phosphonate markers. Magn Reson Med 32:181–188

DiFrancesco D, Noble D 1985 A model of cardiac electrical activity incorporating ionic pumps and concentration changes. Phil Trans R Soc Lond B Biol Sci 307:353–398

Earm YE, Noble D 1990 A model of the single atrial cell: relation between calcium current and calcium release. Proc R Soc Lond B Biol Sci 240:83–96

Honjo H, Boyett M 1992 Correlation between action potential parameters and the size of single sino-atrial node cells isolated from rabbit heart. J Physiol 452:128P

Kodama I, Boyett M 1985 Regional differences in the electrical activity of the rabbit sinus node. Pflugers Arch 404:214–226

Noble D 1991 Ionic mechanisms determining the timing of ventricular repolarisation: significance for cardiac arrhythmias. Ann N Y Acad Sci 644:1–22

Noble D 1995 The development of mathematical models of the heart. Chaos Solitons Fractals 5:321–333

Noble D, Denyer JC, Brown HF, DiFrancesco D 1992 Reciprocal role of the inward currents $i_{b,Na}$ and i_f in controlling and stabilising pacemaker frequency of rabbit sino-atrial node cells. Proc R Soc Lond B Biol Sci 250:199–207

Noble D, Ch'en FF-T, Clarke K, Vaughan-Jones RD 1997 Control of cardiac arrhythmias during calcium overload by regulation of sodium–calcium exchange. Proc XXXIII. IUPS Congress, St Petersburg
Scholz W, Albus U, Lang HJ, Martorana PA, Englert HC, Scholkenes BA 1993 Hoe 694, a new Na^+/H^+ exchange inhibitor and its effects in cardiac ischaemia. Br J Pharmacol 109:562–568
Watano T, Kimura J, Morita T, Nakanishi H 1996 A novel antagonist, No 7943, of the Na^+–Ca^{2+} exchange current in guinea-pig cardiac ventricular cells. Br J Pharmacol 119:555–563
Winslow RL, Jongsma HJ 1995 Role of tissue geometry and spatial localisation of gap junctions in generation of the pacemaker potential. J Physiol 487:126P
Winslow RL, Kimball A, Varghese A, Noble D 1993a Simulating cardiac sinus and atrial networks on the connection machine. Physica D 64:281–298
Winslow RL, Varghese A, Noble D, Adlakha C, Hoythya A 1993b Generation and propagation of triggered activity induced by spatially localised NaK pump inhibition in atrial network models. Proc R Soc Lond B Biol Sci 254:55–61
Wit AL, Janse MJ 1993 The ventricular arrhythmias of ischemia and infarction. Futura Publishing Company, New York

DISCUSSION

Ashmore: This sort of modelling is widespread throughout the neurosciences as well (although of course it reaches its highest form in cardiac physiology!). I would like to underline the point that Denis Noble has made, which is that one way of looking at modelling is as an experimental tool. It is a way of amplifying small effects which are experimentally very hard to get at. From this respect I regard it simply as another weapon in the armoury of the reductionist.

Nagel: It seems to me that you have described something that fits Steven Rose's slogan that wholes constrain parts, but not in a way that's antireductionist. Do you think that the part is constrained by the whole in a way that can be explained by the way in which the whole is constructed out of the parts?

Noble: I strongly believe that we must understand the way in which the whole constrains the parts, as well as understanding the role of the parts in the whole. If I may turn a bit of purple prose on its head, genes can't be 'selfish' because they haven't got that degree of freedom: they are 'captives' of the physiological systems that carry them. This is being deliberately naughty, of course, in order to get across a point, which is that it seems to me that a major complement of a genome project must be to understand the wholes that the genes create well enough to see how they (the genes) in turn are constrained by operating within their environment.

Raff: Have you been able to test your simulations by using multiple electrodes to record from the heart?

Noble: Yes, the sequence of activation in the whole heart corresponds well with the results you get using multiple recording electrodes. I must emphasize that we are totally committed to experimental validation at every stage — I'm not one of those who thinks that one can now let theoretical biology chase free from its experimental base. We validate at every stage, and that includes the simulations of ischaemia, where we're validating against biochemical experiments in which we put whole hearts in the NMR machine and track the metabolites during the sequence of run up to ischaemic arrhythmias.

Raff: How many individual components might you be missing? How many channels and transporters are there, for example, that you still don't know about?

Noble: I've been in this game for about 30 years, so I know very well that in the 1960s I managed to reproduce pacemaker activity in the heart with just four channels and now I'm doing it with about 34. I would be very surprised if they're aren't many more out there that we haven't yet brought in. In fact, that is a question a drug company asked me when we were talking with them about the question of the use of this simulation in future drug discovery. I imagine that there would only be about 10% of the audience who would have been pleased if I said we have probably already got 90–95% of the transporters. The answer most were hoping for is that there are many we haven't yet identified, because they were saying that from their genetic knowledge there must be many more. The point your question is raising is a very important one, which is as you proceed with modelling, how sure are you that you have got a big enough capture of the system (every model is just a partial model of reality) to be confident in what you are doing? The answer is, you can only do that by letting it grow. The other strategy we are adopting is that as more and more channel mechanisms and transport mechanisms are being found, we integrate them in. The virtue of what we do, however, is that when a new chloride channel comes along, for instance, we can at least ask the question about how it behaves in an integrated system.

Kerszberg: This talk and Martin Raff's remarks point to a very different type of reductionism than the one we have been talking about until now. No one is really interested in going deep into a theory of everything in order to understand, say, heart oscillations. Everyone is, however, willing to ignore many channels, molecules and transduction pathways in order to come to an understanding of biological processes such as this.

I'll play devil's advocate for a moment, and say, 'there are so many more ionic channels and each channel you add can change almost everything about the way the model operates, that basically it's a hopeless task'. But I believe this sort of reductionism is what molecular biologists now have to face in a very practical way, because they are each dealing with a particular molecule, receptor or transcription factor in a particular pathway, and they know these molecules are implicated in a variety of other pathways. Nevertheless, they do proceed as if this was not the case. I suspect we are coming to a point where some sort of integration is going to be needed in this respect.

Bray: I wanted to ask the 'Deep Blue' question: that is, to what extent has the computer expanded your understanding of the system? Presumably, when you began the project — starting, as you said with the channels in a single cell — you had some intuitive feeling for what would happen to the ionic balance if you perturbed one of them. Yet, these later integrated systems are enormously more complex and you said that the results were surprisingly counterintuitive. Do you feel that you *plus* the computer together have an expanded understanding of the system?

Noble: That is a good way of putting it, because using computers is like having vastly more thinking power at your finger tips, remembering of course that the computer is only doing what you're putting in by the way of thinking earlier on. It is the speed of the machine that enables some of these results to come out. I could ask the question: would inhibition of sodium–calcium exchange be counterproductive in a whole heart situation in the complex arrangements that occur when cells go ischaemic? But without this kind of simulation I wouldn't know how to answer the question. There are things here that we could not conceivably do without the computer.

Williams: This comes back to what I was saying earlier. If you have a phase transition, complex cooperativity of one kind is introduced into the system. Now in some contexts, cooperativity has been quite easy to understand. For instance, if you ask the question as to why sodium chloride forms a crystal at room temperature whereas nitrogen does not, you can solve this all the way down to the level of simple additive equations for interactions between atoms. When you are explaining these systems, you use an *additive* cooperativity between a large number of different atoms. Cooperativity at a phase transition is quite different and is not simply additive between atoms so that the explanation lies in complexity at a mathematical level. A bacterial cell is monitoring itself in a different cooperative way, employing about 20 different chemical elements in flow. This 'cooperativity' is due now to kinetic feedback and is again mathematically complex, so that there could be many solutions due to many variables but we see only one when we look at a given organism. I do not believe that we can sort out the reason for the observed survival of the bacteria by referring to isolated atoms or molecules. This seems to parallel your treatment of the heart. You seek for a behavioural pattern but if you adjust input you may get very different answers, of which you see only one. Do you just fiddle the rate constants or do you allow the computer to find those constants, so you have actually meaningful constants which represent the cooperativity?

Noble: The interrelations between them are signalled by the processes that the cells are known to use, such as intracellular calcium and voltage, and that enable the transporters to talk to one another.

Williams: So do you know the feedback constants as well as the input signals?

Noble: We don't know everything. Obviously, in any such modelling there are problems with the experimental data. As you build up and up there are increasing degrees of freedom. This is a dilemma. When we get an important result such as the sodium–calcium exchange result, we have tried to resolve that problem by proceeding to pan through the ranges of parameters that we think that some of the crucial parts of the interacting system could have to test the robustness of that particular simulation. I think this is the best we can do at the moment.

Williams: In a bacterial cell, for example, if you look at the ferrous iron level, you will find it is coupled to many other things, just as in the heart you have shown that the calcium level is coupled to a large number of other parameters. The understanding of this is what I think of as belonging to the machine language world, where everything is working together to make a cooperative flowing whole. The problem is the underlying complexity. Function in machines and in biological systems is complex.

Bateson: How does your theoretical activity relate to that of mathematicians like Ian Stewart, who bring analytical equations from chaos theory to bear on the heart? They obtain counterintuitive results similar to yours. It seemed to me, though, that the parameters put into their equations do not have any biological reality. Is there a point of contact between your approach and theirs?

Noble: Yes, I think they're complementary. I see it in exactly the same way with regard to people such as Winfree (1987), who first demonstrated the importance of scroll and spiral waves. It seems to me that on the one hand you've got modelling like that, which is asking some questions about systems of equations that are sufficiently simple to do some analytical work on, and on the other hand there are our systems which are highly concrete: they're founded on going down to the lowest level we think is necessary. We depend on absolutely massive computing power to do this. It seems to me that you need both approaches, because the dilemma of our kind of approach is that when you arrive at the result you may not know how generic it is. Then you need the analysis of simpler systems that can attack the same questions but with a greater degree of generality so that the mathematical results that come out are generic rather than specific. But if you just have the generic type of results the question you would naturally ask is, how is that reaching down to the real biology underneath it all? That's why Lewis Wolpert calls me a reductionist.

Wolpert: With your computer simulations of cardiac physiology, do you think you've understood what's going on? This is getting back to what we mean by understanding.

Noble: Not at all. One of the interesting things is that there is a lot there that I'm well aware we don't understand.

Wolpert: Are you saying that you don't really have an understanding as to why what has happened has happened?

Noble: Let me be specific about two of the things I showed. I think we can give a pretty good account now of why it turns out that up-regulating the sodium–calcium exchanger produced the effects we saw. We have been able to do this because in that case the computer simulations have led us along and our intuitions have too. We can work through the network of interaction and say 'now that we understand how all this is interacting and given this kind of modelling, we can now see why we should be doing this or that to achieve particular objectives'. With regard to the question of why it is with many of the perturbations that occur in abnormal arrhythmia that we end up with these wandering, chaotic spirals, in one sense one understands this from the work of Winfree (1987) and others using complexity theory and simpler models, but I don't feel entirely happy with this type of explanation and I'm still puzzling over it. If I can talk now about happiness, the answer to the question that would make me happy concerns whether I can understand why it is that the heart is so structured that these arrhythmias do *not* happen normally. This really would be exciting.

Reference

Winfree AT 1987 When time breaks down. Princeton University Press, Princeton, NJ

General discussion

Is all science reductionist?

Wolpert: I want to put to you the proposition that there is no science that is not reductionist. What I mean here by 'reductionism' is that one takes a set of phenomena and uses a small set of concepts to explain the observations. In these terms, Newtonian mechanics is reductionist. What you can or can't explain is a separate issue. I'm asserting that there is no science that is not intrinsically reductionist.

Gray: Before I try to give some counter-examples, I think we have to be clear— counter-examples to what? This is not as clear as perhaps you think.

Thomas Nagel gave us a very good introduction this morning to the substantive philosophy of reductionism, with his distinction between epistemological and ontological antireductionism. But there's also a sociology of reduction and antireduction. There is a lot of surplus meaning attached to this issue, and your question has a great deal of it. It has a kind of passion and subtext that has very little to do with the perfectly clear scientific distinctions that Denis Noble was making in his paper (Noble 1998, this volume). It doesn't matter whether one calls what Denis has done 'reductionist' or 'non-reductionist', the way the elements of the system fit together is perfectly clear. To me, the only useful way of putting the question you are asking would be to say something like this: you have reductionism in science when you account for concepts at one level (and I don't think you can do this without the word level somewhere in it) by reference to elements at a lower level. That means you have to start by thinking in terms of higher and lower levels. The standard sociology of science is that you start with physics and you move to chemistry, then biology, then behaviour, and perhaps then to social systems. If I am to give you a counter-example, somewhere in that hierarchy there would have to be a successful piece of explanation in science where the elements in the explanation are not at a lower level than the phenomena you are trying to explain.

If you consider the way the genetic code works, we get a great deal of mileage in understanding by talking about it in terms of information. A standard way in which people look through DNA sequences is to seek particular kinds of repetition of elements: it doesn't matter in the least, when they do that, that the letters in the sequences that they're using are names of particular pieces of chemical. That is more than a metaphor, because it is actually what is going on: DNA sequences are being used as information transmitted from generation to generation to specify the construction of certain organisms. The level of information is not one which can be reduced to the level of chemistry, because that same level can be instantiated in many other ways — it doesn't have to be instantiated in the sequence of bases. The sequences of bases do not themselves force the concepts of information that are used at that level of discourse. To me, this is an example of a successful approach to a problem in science which is not reductionist. It is almost antireductionist in the sense that you are going from the bottom level up to a top level.

Quinn: I think there are two kinds of science that are not reductionist. One of them is taxonomy: categorizing things without exactly knowing why that's so. Linnaeus advanced science in this way. The other one is picking what you think might be a reasonable number without a hypothesis and measuring. High temperature superconductivity people at one level are smart, and at another level it's pure luck. Some of the best science is unpredictable and is done by blind luck. If you do try a reductionist science which most of us do most of the time, and you don't succeed, I think it's really important to know when you can't do it. The aftermath of this ought to be honesty, and then you move onto something else or you try something else. There are things that at a given time in history are not amenable to reasonable reductionism.

Garcia-Bellido: It is reductionism when you are making propositions or explanations to elements in the next level down. It is antireductionism when you explain things by properties of higher levels to the one you want to study. This has happened throughout the history of science. The global explanation is the opposite of reductionism, but both are science. Lewis Wolpert, your proposition of positional information is a global proposition to properties which are underneath. You are not explaining the properties of cells by explaining what the subcellular components are doing, but you are describing a general property as a rule, which has to embrace observations at another level to confirm it.

Mitchison: I haven't read very much of Linnaeus, but my impression was that when he wasn't naming small ill-smelling Compositae after his colleagues, he was actually often using notions such as the numbers of stamens (which, incidentally, he rendered in splendid language of the bedchamber). So he was striving to carry out a sort of counting operation, and of course modern numerical taxonomy does just this with DNA or protein sequences. This is where the notion of information comes in. When people these days try to give a precise measure of information and distances between protein sequences and so on, what they're doing is converting data into probabilities which are directly related to the type of environment that that residue finds itself in a protein. It is extremely closely connected with the components that you are looking at. So I don't think either of those are terribly convincing examples: they may have an historical residue of non-reductionism but they're certainly pretty reductionist nowadays.

Raff: If you had re-stated your proposition to say that experimental science is almost always reductionist, I don't think many people would object.

Kerszberg: I disagree. It must be recognized that thermodynamics, for example, is the science of saying many important things about the system without knowing anything about its internal workings. Computer science, one of the most important sciences together with biology today, is the science of saying something about computing machines without worrying in the least about the hardware. It is only the logical structure of the machine that is considered.

Quinn: That's reduction.

Kerszberg: No, it's not reduction: it's a holistic approach from the beginning.

Wolpert: Isn't thermodynamics reductionist? It is couched in terms of energy, and that is a reductionist concept.

Williams: Let us take a common statement in thermodynamics and ask whether we understand it. If you make the statement of the second law saying 'the entropy of the universe is increasing', it is difficult to know what that means. To know the entropy you have to know the statistics of the system. I learned a strong lesson from somebody about this problem, when he said to me 'have you counted the number of photons in the problem?' I had only considered material particles. The next stage down is to ask 'have you tried to count the number of gravitons?' This is ludicrous, because we don't even know that the graviton exists: it's a theoretical idea. Thermodynamics is not a secure basis for general thinking about reduction. It deals with systems.

Rose: A couple of quick points. First, I don't think it is the case that because you can describe a complex set of phenomena by mathematical equations then that is, by definition, reductionism. Secondly, I think there's a distinction between methodological reductionism, which is the experimental programme that Martin Raff is talking about (i.e. that we need to keep variables steady, we need to control parameters, etc.) which is how most of us do productive science in the lab, and the sort of philosophical reductionism that Thomas Nagel described this morning. I still maintain that while reductionism is essential, it cannot answer certain sorts of questions because certain properties only emerge at a particular level. It cannot answer the question about why the frog is jumping by telling us about the composition of actin, unless you know that there's a snake which is about to swallow the frog. That is a crucial component of the sort of explanations we need in biology.

Reference

Noble D 1998 Reduction and integration in understanding the heart. In: The limits of reductionism in biology. Wiley, Chichester (Novartis Found Symp 213) p 56–72

Muscle contraction

K. C. Holmes

Department of Biophysics, Max Planck Institute for Medical Research, Jahnstrasse 29, 69120 Heidelberg, Germany

Abstract. Understanding muscle contraction goes to the heart of one of the fundamental questions posed by classical philosophy, namely the nature of the *πνευμα ψυχικον*. The nature of 'understanding' has altered greatly during the last two millenia, particularly in response to the development of the concept of energy. Moreover, understanding contraction depends on understanding muscle structure. Galen was the first to make a detailed anatomical examination of the mode of action of muscles and recognized the heart as a muscle, but this line of research was not pursued until Leonardo da Vinci rediscovered it 1400 years later. Vesalius used the phrase *Machina Carnis*, but it was first Descartes who proposed a neuromuscular machine. However, the level of understanding of the physiology of muscle depends critically on the resolution of the available anatomy. Radical new insight was provided by electron microscopy. But an understanding at a physicochemical level is only possible if the structures of the components are known at atomic resolution. These have become known in the last five years and have led to dramatic progress. The present level of understanding of muscle is a physicochemical explanation of how the hydrolysis of ATP by the component proteins actin and myosin leads to movement.

1998 The limits of reductionism in biology. Wiley, Chichester (Novartis Foundation Symposium 213) p 76–92

Prelude

Understanding the physicochemical basis of muscle contraction goes to the heart of one of the fundamental questions posed by classical philosophy, namely the nature of the *πνευμα ψυχικον* — *spiritus animalis*.

In the beginning of the 2nd century Galen made a detailed anatomical examination of muscles. In the ensuing millenium Galen's work was not extended and even Galen's insight that muscles pull rather than push seems to have been forgotten, since at the beginning of the 16th century, on the basis of his own anatomical examinations Leonardo da Vinci wrote: '*perchè l'ufizio del musculo è di tirare e non di spingere*'. A few years later Vesalius used the phrase *Machina Carnis* to underline the fact that the origin of force in muscle was the flesh itself, and Descartes proposed a neuromuscular machine: the nerves carry a fluid from the pineal gland to the muscle which make them swell and shorten. Erasistratus of the Alexandrian school had in fact produced a similar theory 1800 years earlier. Descartes took over the view held by the Greeks that muscles increase their volume as they shorten. A little later, Swammerdam was to show that muscles contract at constant volume, which invalidated this whole class of pneumatic theories. However, other mechanical models were soon proposed. Alongside such mechanical thinking, however, vitalism survived into the 19th century, and one needed the whole fabric of metabolic biochemistry and thermodynamics to support the concept that muscle is a chemical machine driven by isothermal combustion.

Our understanding of the physiology of muscle depends critically on the resolution of the available anatomy. Some early insight was provided by light microscopy. However, the first radical new insight was provided by electron microscopy. Ultimately, an understanding in physicochemical terms is only possible if the structures of the components are known at atomic resolution. These have become known in the last five years and now allow us to describe in some detail how the hydrolysis of ATP by the component proteins actin and myosin leads to movement. The understanding of muscle contraction is an important example of the success of the reductionist method.

The contraction of voluntary muscles in all animals takes place by the mutual sliding of two sets of interdigitating filaments: thick (containing the protein myosin) and thin (containing the protein actin) organized in repeating units called sarcomeres each a few microns long. The sarcomeres give skeletal muscle its characteristic crosss-striated appearance first seen by van Leeuwenhoek. The relative sliding of thick and thin filaments is brought about by 'cross-bridges', parts of the myosin molecules which stick out from the myosin filaments and interact cyclically with the thin filaments, transporting them by a kind of rowing action. During the process ATP is hydrolysed to ADP, which provides the energy.

The protein myosin was discovered more than 100 years ago by Kühne (1864). Von Helmholtz and later A.V. Hill (1922) showed that as muscle does work it releases heat proportionately thereby strengthening the concept of muscle as a chemical machine. ATP was discovered by Lohmann (1931) and Lyubimova & Engelhardt (1939) showed that the ATP was hydrolysed by myosin and is the immediate energy source for muscle. Working with Albert Szent-Gyorgyi in Szeged, Straub discovered that 'myosin' was actually two proteins, myosin and actin (Straub 1943). ATP was also shown to be a 'relaxing factor'—i.e. ATP also dissociates actin and myosin. Moreover, Szent-Gyorgyi was able to show that glygerol-treated muscle fibres, containing only actin and myosin, shorten on the addition of ATP. The dichotomy of the action of ATP (a relaxing factor that drives contraction) remained an enigma to be explained later by Lymn & Taylor (1971). In the meantime, the myosin molecule was characterized and was shown to consist of two heavy chains and two light chains. A soluble proteolytic fragment of myosin, heavy mero-myosin (which contains the globular 'heads' of myosin) contains the ATPase activity (Szent-Gyorgyi 1953), the rest of the molecule forming a long α-helical coiled-coil. The ATPase activity was later shown to reside in the 'head' itself (Margossian & Lowey 1973a,b) (often called S1) which constitutes the morphological 'cross-bridge'.

The first molecular theories, which appeared in the 1930s, were derived from polymer science. They proposed that there was a rubber-like shortening of myosin filaments brought about by altering the state of ionization of the myosin. This aberration was corrected by the seminal works of H. E. Huxley (Huxley & Hanson 1954) and A. F. Huxley (Huxley & Niedergerke 1954) which showed that sarcomeres contained two sets of filaments which glided over each other without altering their length. The question naturally arose: what made them glide? The myosin cross-bridges were discovered by electron microscopy (Huxley 1957, 1958) and subsequently shown both to be the site of the ATPase and also to be the motor elements producing force and movement between the filaments. Two conformations of the cross-bridge could be detected in insect flight muscle (Reedy et al 1965). Progress was then rapid so that at a historic Cold Spring Harbor Symposium in 1972 the outline of the molecular mechanism of muscle contraction could be enunciated. The cross-bridge was thought to bind to actin in an initial (90°) conformation, go over to an angled (45°) conformation and then release (Huxley 1969, Lymn & Taylor 1971). For each cycle of activity one ATP molecule would be hydrolysed. The actual movement could be measured by physiological experiments on contracting muscle and was shown to be about 80–100 Å (Huxley & Simmons 1971). Since the cross-bridge was known to be an elongated structure, such a distance could be accommodated by a rotating or swinging crossbridge model (Fig. 1) .

The cross-bridge cycle

In the absence of ATP, the myosin cross-bridge binds tightly to actin filaments. However, it also binds and hydrolyses ATP. ATP binding brings about a rapid dissociation of the cross-bridge from actin. Thus the cross-bridge can bind either actin or ATP but to both only transiently. The presence of the ATP γ-phosphate is crucial for dissociation since ADP alone has little effect. Solution kinetic observations were very important in establishing the relationship between the hydrolysis of ATP and the generation of force. A key feature of this process is the observation that transduction of the chemical energy released by the hydrolysis of ATP into directed mechanical force should occur during product release (ADP and inorganic phosphate, P_i) rather than during the hydrolysis step

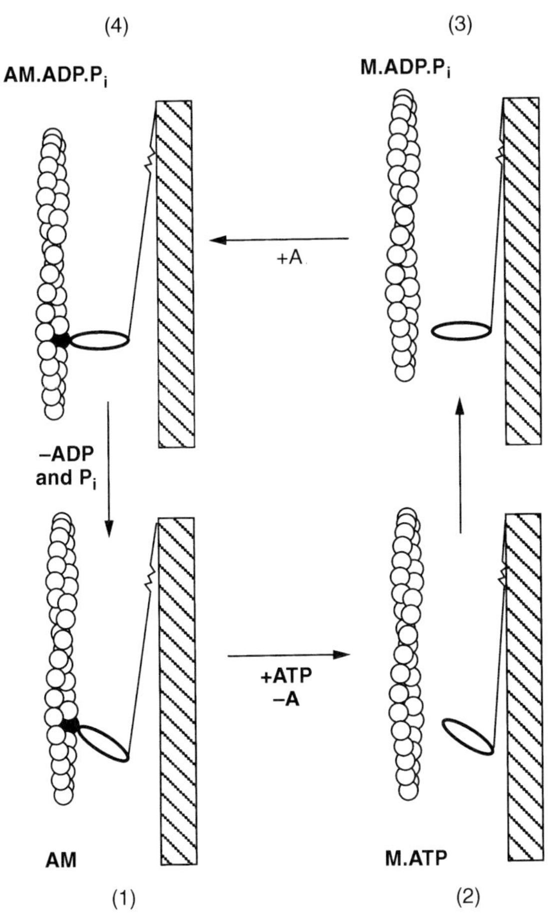

FIG. 1. The Lymn–Taylor cycle (Lymn & Taylor 1971): the myosin cross-bridge is bound to actin in rigor 45° position — 'down' (1). ATP binds which leads to very fast dissociation from actin (2). The hydrolysis of ATP to ADP and P_i leads to a return of the myosin cross-bridge to the 90° 'up' position whereupon it rebinds to actin (3). This leads to release of the products and return to (1). In the last step actin is 'rowed' past myosin.

itself (Lymn & Taylor 1971). Without actin, myosin is product-inhibited and is a poor ATPase.

Mg-ATP rapidly dissociates the actomyosin complex on binding to the ATPase site of myosin; myosin then hydrolyses ATP and forms a stable myosin–products complex; actin recombines with this complex and dissociates the products, initially the γ-phosphate ion, thereby forming the original actin–myosin complex. Force is generated during the last step.

Although the swinging myosin cross-bridge hypothesis of muscle contraction had become the textbook norm by 1972, it has proved remarkably difficult to catch a bridge *in flagranti delicto* (see review by Cooke 1986). Nevertheless, the swinging cross-bridge hypothesis provides by far the best framework for correlating and explaining the vast muscle literature. However, there were indeed problems. In fact the hypothesis was never very clear about how the cross-bridge moved on actin and had been modified over the years into a swinging lever arm hypothesis in which the bulk of the cross-bridge is envisaged to bind to actin with a more or less fixed geometry and only the distal (C-terminal) part of the myosin molecule moves (Holmes 1997) (Fig. 2). A swinging lever arm explains why substantial changes in the cross-bridge orientation were not visible: only a small fraction of the cross-bridge mass moves. Furthermore, it gradually became clear that the proportion of cross-bridges taking part in a contraction at any one time was only a small fraction of the total, making the registration of active cross-bridge movement doubly difficult.

High resolution structures yield insight

The atomic structures of actin and myosin provided the *rinascimento* for muscle research. In particular, the crystal structure of the myosin subfragment 1 (Rayment et al 1993b) endowed the myosin cross-bridge with an extended C-terminal tail which looked like a lever arm and, moreover, a lever arm which was in the correct orientation and position to function as a lever arm (Rayment et al 1993a). In the last year a number of independent experiments have provided results which are in excellent accord with the idea that the C-terminal tail functions as a lever arm and indeed provide evidence that it can move. Furthermore, new crystal structures (Fisher et al 1995, Smith & Rayment 1996b) with analogs of ADP.P_i bound appear to show an alternative orientation of the lever of the anticipated kind (Holmes 1996).

Atomic structures of actin and myosin

Actin (thin filament) fibres are helical polymers of g-actin (globular-actin). The structure of monomeric actin, which contains 365 residues and has an M_r of 42 000, was solved by protein crystallography as a 1 : 1 complex with the enzyme DNase I (Kabsch et al 1990). Orientated gels of actin fibres (f-actin), a helical copolymer of actin which has 13 molecules in six turns repeating every 360 Å, yield X-ray fibre diagrams to about 6 Å resolution. It was possible to determine the orientation of the g-actin monomer which best accounted for the f-actin fibre diagram (Holmes et al 1990) and thus arrive at an atomic model of the actin filament (Lorenz et al 1993).

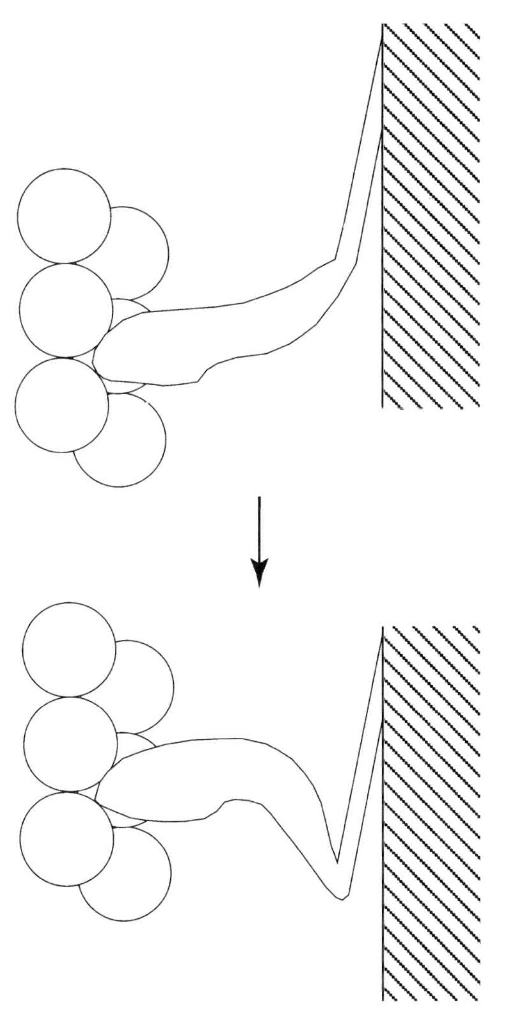

FIG. 2. Numerous experiments (mostly negative) indicated that the scheme shown in Fig. 1 needed revision: only the distal part of the cross-bridge moves (Cooke 1986).

The cross-bridges comprise a part of the myosin molecule, namely subfragment-1 of heavy mero-myosin (S1). The structure of chicken S1 has been solved by X-ray crystallography (Rayment et al 1993b). (In the following, references to residues in the chicken structure will be prefaced with 'gg'.) This study shows the S1 (which has 884 residues) to be tadpole-like in form, with an elongated head, containing a seven-stranded β-sheet and numerous associated α-helices forming a deep cleft, with the actin binding sites and nucleotide binding sites on opposite sides of the sheet. The cleft separates two parts of the molecule which are referred to as the 50K upper and 50K lower domains. The C-terminal tail, sometimes called the 'neck' and which also provides the connection to the thick filament, forms an extended α-helix which binds two light chains. The ATP binding site contains the typical 'P-loop' motif which is also found in the G proteins.

By fitting the atomic structures of f-actin and S1 into 3D cryoelectron microscope reconstructions, one arrives at an atomic model of the actin myosin complex (Rayment et al 1993a) (Plate 1a). In particular, this model establishes the spatial orientation of the S1 myosin fragment in the active complex. For example, one finds that the cleft in myosin extends from the ATP binding site to the actin binding site and that the opening and closing of this cleft is very likely to provide the communication between the ATP site and the actin binding site. The actin binding site spans the 50K upper and lower domains and the ATP binding site extends from the 50K upper domain into the 50K lower domain. Furthermore, the very extended C-terminal α-helical neck of S1 is ideally placed to be a lever arm. The lever arm joins onto the bulk of the molecule via a small compact 'converter domain' (Houdusse & Cohen 1996) which lies just distal to a broken α-helix containing two reactive thiol groups known as SH1 and SH2. Numerous experiments point to the putative 'hinge' for the lever arm being in the SH1–SH2 region of the molecule (see Holmes 1997 for review).

Movement of the lever arm

The cross-bridge movement was studied in time-resolved experiments of contracting frog muscle using low-angle X-ray fibre diffraction (Huxley et al 1981, Irving et al 1992). While these results are fully consistent with the swinging cross-bridge theory, the complexity of the system and low resolution of the method precludes an unambiguous interpretation. Therefore a simpler model system has been studied: 'decorated actin', which is an actin filament with a myosin cross-bridge bound to each actin. Whereas the structure of decorated actin in the rigor state (no bound nucleotide) has been extensively studied (Moore et al 1970, Milligan & Flicker 1987), corresponding studies in the presence of ATP are difficult since the binding of ATP leads to rapid dissociation of the cross-bridges from actin. Time-resolved electron micrograph studies in fact show no bulk change of the cross-bridge orientation on binding ATP before dissociation takes place (Pollard et al 1993), whereby a reorientation of the lever arm would not have been detected at the resolution attainable. High-resolution electron micrographs of actin decorated with smooth muscle myosin, however, show a 30–35° rotation of the lever arm on binding ADP (Jontes et al 1995, Whittaker et al 1995). Although the main movement of the lever arm would be expected to be associated with phosphate release, since this is a step associated with a large change in free energy, some fraction of the movement could arise from ADP binding and release. Moreover, this movement should be recoverable on adding ADP to actomyosin, which indeed it is. Although the effect has only been found in smooth muscle myosins it is generally important in providing the first direct demonstration of a nucleotide-induced lever-arm swing. More recent work using a spin label attached to a light chain also supports this result (Gollub et al 1996). Moreover, X-ray diffraction studies of muscle fibres loaded with exogenous smooth muscle cross-bridges show the predicted changes in the fibre diffraction pattern resulting from the lever arm swing on binding ADP (Poole et al 1998).

Purified myosin cross-bridges (S1) can be attached to a substrate and used to transport actin filaments *in vitro* in the presence of ATP. A study by Uyeda et al (1996) shows that the speed of actin transport in motility assays is proportional to the length of the lever arm. Moreover, the fulcrum appears to lie near the broken helix (gg690–710) which contains the especially reactive thiols (SH1, SH2) of myosin. A similar result has been obtained by Manstein and co-workers using a 'synthetic' lever arm made from repeating α-actinin repeats in place of the light chain binding region (Anson et al 1996).

Mutagenesis studies also indicate the importance of the SH1–SH2 region in controlling lever arm movement (Patterson & Spudich 1996). A ggG699A mutation, between the SH1 and SH2 groups, slows myosin transport of actin 100-fold (Kinose et al 1996).

Specific fluorescent markers attached to the 'regulatory' light chain show a small angular movement on contraction (Allen et al 1995, Irving et al 1995), whereas the 'lever arm hypothesis' expects about 60° rotation. However, if only a fraction (c. 10–20%) of the cross-bridges in active muscle take part in contraction at any one time, the magnitude of this apparent rotation can be proportionally scaled up towards the anticipated value.

The myosin cross-bridge has two conformations

According to the Lymn–Taylor scheme (Fig. 1) the myosin cross-bridge would be expected to have two discernible conformations: (1) when it first attaches to actin with the products of hydrolysis still bound and the lever at the beginning of the working stroke; and (2) at the end of the working stroke when the phosphate and ADP are released. This sequence is often referred to as the 'power stroke'. The end state is referred to as 'rigor', since it is the state muscle enters on ATP depletion. It is also called 'strong' because it binds to actin quite tightly. The initial state is called the 'weak binding state' because of its low affinity for actin (see Geeves & Conibear 1995). We might anticipate that these two states of the myosin cross-bridge might exist independently from actin and indeed protein crystallography shows this to be the case.

The chicken S1 structure was solved without bound nucleotide. Furthermore, the chicken S1 crystal structure fits excellently into the electron micrograph reconstructions of the strong actin–myosin nucleotide-free interaction (decorated actin) . Therefore the crystal structure of chicken S1 would appear to represent the end of the power stroke or rigor state.

In addition, Rayment's laboratory have studied a crystalline fragment of the *Dictyostelium* myosin II cross-bridge which has been truncated after residue 761 (equivalent to gg781). The truncation eliminates the lever arm and the associated light chains (which aids crystallization). However, the converter domain is still present. The crystal structures of the 761 construct have been determined with a number of ATP analogues, particularly ADP.BeFx (Fisher et al 1995) and ADP.vanadate (Smith & Rayment 1996b). ADP.vanadate complexes are used as analogues of the transition state or possibly of the ADP.P_i state. While the ADP.BeFx state looks similar to rigor, the ADP.vanadate structure shows, compared with the chicken rigor structure, dramatic changes in shape of the S1 structure, There is a closing of the 50K upper/lower domain cleft, particularly around the γ-phosphate binding pocket, and large movements in the C-terminal region. The 50K upper/lower domains rotate a few degrees with relation to each other around the helix gg648–666 in a way which closes the nucleotide binding pocket (Plate 1b) — a movement of some 5 Å. At the same time the outer end of the long helix (the so-called switch 2 helix, residues gg475–507) bends out 24° at residue V497. This is coupled to a rotation of the converter domain (711–781) by 70° (Houdusse & Cohen 1996). The fulcrum is provided by the mutual rotation of the distal part of the SH1–SH2 helix around the distal part of the switch 2 helix.

A model of this new state is shown in Plate 2a. The coordinates of the missing lever arm have been generated from the chicken coordinates by superimposing the converter domains. The light chains have been omitted for clarity. For comparison (Plate 2b) we have generated the corresponding diagram from the coordinates with ADP.BeFx bound in the active site (Fisher et al 1995) which, for reasons stated below, we take to be the ADP state. On comparing Plates 2a and 2b one sees that the end of the lever arm has moved through 12 nm along the actin helix axis, which is greater than most estimates of the size of the power stroke. Therefore, it would appear that the ADP.vanadate state is indeed a model of the anticipated 'beginning of power stroke' state.

The mechanism for coupling the movement of the lever arm with the status of the nucleotide binding pocket revealed by this structure suggests that the two events are tightly coupled: pocket closed, lever up (beginning); pocket open, lever down (end).

For ATP hydrolysis the 50K upper/lower domain cleft should close

Smith & Rayment (1996a) point out the similarity of the active site of myosin in the closed form with Ras and other G proteins and that the enzymes probably share a common mechanism for hydrolysis (Schweins et al 1995). The differences between the open and closed forms of the myosin cross-bridge in the neighbourhood of the active site reside almost entirely in the conformation of the linker region (gg465–470) which joins the 50K upper and lower domains. Smith and Rayment point out that this region is structurally equivalent to the so-called switch 2 region in Ras with which it also has a very strong sequence homology. The mutual rotation and closing of the 50K upper/lower domains cleft causes this region to move by about 5 Å. In the chicken crystal structure (open), which has no bound nucleotide and should therefore be in the rigor conformation, the switch 2 region is pulled away from the nucleotide binding pocket. A similar movement of the switch 2 region depending on whether di- or trinucleotide is bound is also found in the G proteins. Only in the closed form (ADP.vanadate) can the hydrogen bond between the carbonyl of ggG466 and the γ-phosphate (Plate 1b), which is an invariant characteristic of the G protein active sites, be formed. Because of the importance of ggG466 (and other residues) for hydrolysis it is difficult to see how hydrolysis can proceed in the open (rigor) form which would therefore appear not to be an MgATPase: the closing would appear to be essential for hydrolysis.

ADP.BeFx can produce both open and closed states

ADP.BeFx is thought to be an analogue for ATP. Fisher et al (1995) solved the structure of *Dictyostelium* S1 truncated at 761 with ADP.BeFx bound in the active site and found it to be remarkably similar to chicken S1 without nucleotide. This result appears to show that the structure of the ATP state is 'open', which is puzzling since it would not be able to hydrolyse the ATP. Moreover, the attitude of the converter domain (and hence the neck) is close to rigor which is also unexpected for the ATP state (cf. Plate 2b with Plate 1a). More recently Schlichting et al (1998) have solved the structure of an ADP.BeFx complex of truncated S1 and find it to be essentially identical to the ADP.vanadate complex. The active site is closed and the converter domain is in the rotated configuration. The construct used in this case was truncated at position 754 and is therefore seven residues shorter than that used by Fisher et al (1995). This results in a tighter binding of ADP (Kurzawa et al 1997). Apparently, on account of this difference, in the shorter construct the binding energy of ADP.BeFx is adequate to tip the scales for the closed structure, whereas in the longer construct it was not. Therefore one can picture the transition between the two forms of myosin as being sensitively poised (but well determined — intermediate states have yet to show up): the structure solved by Fisher et al apparently corresponds to the ADP-bound state whereas the structure solved by Schlichting et al corresponds to the ATP-bound state. Comparing the Fisher et al (1995) ADP.BeFx state (Plate 2b) with chicken rigor (Plate 1a) there is in fact a 10° movement of the lever arm. While this may reflect small changes in the angle of the lever arm induced by the binding of ADP it could also reflect species differences.

Phosphate release: actin binds to the open form of the 50K upper/lower cleft and thereby facilitates phosphate release

The closed structure found with the ADP.vanadate generates a tight hydrogen bonding pattern for the γ-phosphate which probably explains the high phosphate affinity. This interaction in turn is important for stabilizing the closed form. Opening the cleft destroys the γ-phosphate binding pocket. Energy-filter cryo-electron microscopy of decorated actin (Schroeder et al 1998) shows that the cleft may be open in the actin–myosin complex. Therefore it seems very likely that actin binding opens the cleft rather than closes the cleft as was suggested earlier (Rayment et al 1993a). Opening the cleft destroys the phosphate binding site and facilitates γ-phosphate release (a 'back door enzyme'; Yount et al 1995).

Although kinetic studies provide evidence that the actin–myosin binding in the presence of nucleotide is a multi-step process, there are no structural data on an initial weak binding of the closed form to actin. A consistent scheme may be developed by postulating that there is an additional transitory state, a bent closed form. We suppose that actin binds myosin with one main set of contacts at approximately constant geometry, namely as is seen in the rigor actin–myosin complex (i.e. the open form of myosin). The 50K lower domain probably forms the invariant contacts to actin: the switch from weak to strong binding probably involves the recruitment of loops (the 50K–20K loop and the '404' loop) from the 50K upper domain to form the strong binding state. When confronted with myosin in the closed form actin probably binds the 50K lower domain first, which binds actin weakly. The subsequent binding of the loops produces an open form which releases the γ-phosphate and binds actin strongly.

Summary

Crystallographic studies show two distinct structural states for myosin S1: the 'open' or 'end' conformation which is characterized by the absence of nucleotide (rigor); and the 'closed' or 'beginning' state, which is favoured by binding ATP or the products complex (ADP.P_i). Myosin transports actin by switching between these two states. 'Open' and 'closed' refer to the status of the ATP binding site

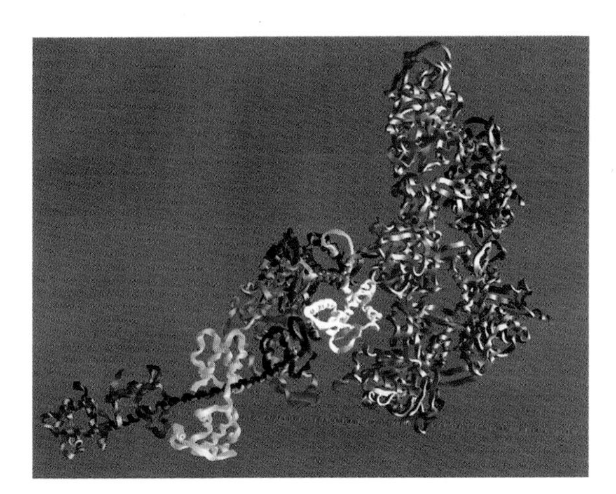

PLATE 1a The structure of the actin–myosin complex (Rayment et al 1993a, Schroeder et al 1993): shown are (right) five actin molecules in an actin helix (Holmes et al 1990) and (left) a myosin cross-bridge (S1) (Rayment et al 1993b). Shown are: 25K fragment (green); 50K upper fragment (red); 50K lower fragment (white); the disordered chain between the 50K domain and the 20K domain is shown as a yellow loop: note this loop has been modelled; the first part of the 20K domain including the SH2 helix (until 699) is shown in light blue; the SH1 helix, converter domain and the C-terminal helix (the 'neck') are dark blue; the regulatory light chain is magenta; and the essential light chain is yellow. Plates prepared with GRASP (Nicholls et al 1991).

PLATE 1b A view of the ATP binding site looking out from the actin helix. Shown are: the P-loop (green); an MgATP molecule with the base at the back and the three phosphate groups in the front (carbon, yellow; nitrogen, blue; phosphate, light-blue; oxygen, red; magnesium, green); parts of the 50K upper domain (red) including the so called 'switch 1' region (right); the switch 2 region in the open ADP (white) and closed ATP (grey) conformations – the conserved glycine (gg466) is shown in grey or white – note that this residue moves about 5 Å between the two conformations; and in blue the helix (gg648–666), which acts as fulcrum for the relative rotation of the 50K upper and lower domains.

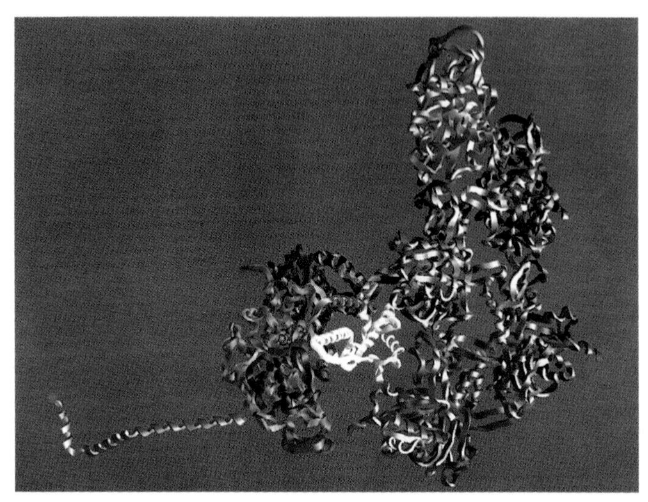

PLATE 2a The 'end' state: the rigor complex (as in Plate 1) modelled from the crystallographic data on the *Dictyostelium* myosin motor domain truncated at residue 761 and complexed with ADP.BeFx (Fisher et al 1995). To establish the orientation w.r.t. the actin helix (right) the 50K upper and lower domains have been superimposed on the corresponding domains in the rigor structure shown in Plate 1. Although the motor domain has bound nucleotide, it appears to be close to the rigor state. The missing 'neck' region or lever arm (light blue) has been modelled from the chicken S1 data by superimposing the converter domains. The light chains have been omitted for clarity.

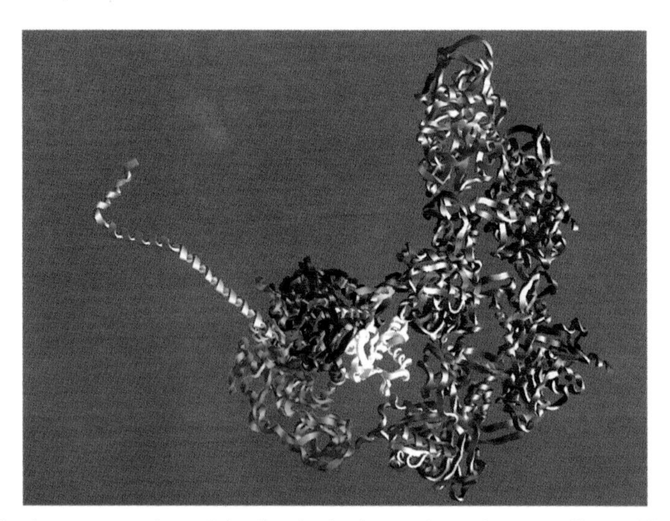

PLATE 2b A reconstruction of the 'beginning' state from the crystallographic data on the *Dictyostelium* construct truncated at 761 and complexed with ADP.vanadate (Smith & Rayment 1996b). Note the 70° rotation of the converter domain. The missing 'neck' or lever arm has been modelled from chicken S1 data by superimposing the converter domains. The light chains have been omitted for clarity. The rotation of the converter domain is controlled by the bending out of the 'switch-2' helix. The end of the lever arm moves about 12 nm between the two states.

which extends from the 50K upper domain across to the 50K lower domain. This in turn is coupled to the rotation of a C-terminal lever arm. In the closed form the lever arm is at the beginning of the power stroke whereas in the open form it is at the end of the power stoke. The preference for open or closed is also controlled by binding to actin. We hypothesize that the closed state binds only weakly to actin. On this basis we can correlate the structural states with the Lymn–Taylor cycle.

Starting from an actin–myosin complex at the end of the power stroke, the binding of ATP brings about rapid closure of the cleft and concomitant release from actin. The closed state hydrolyses ATP to ADP.P_i without attaching to actin. Thereafter, the rebinding of myosin in the closed or 'beginning' conformation of the products complex to actin opens the cleft to facilitate release of the γ-phosphate. Release of phosphate induces an isomerization to the open 'end' conformation since it is the presence of the γ-phosphate which stabilizes the closed form. The isomerization results in large changes of angle of the 'lever arm' (at the distal part of the myosin head). Since the S1 is strongly attached to actin at this stage this results in a 12 nm transport of actin past myosin.

References

Allen TS, Sabido-David C, Ling N, Irving M, Goldman YE 1995 Transients of fluorescence polarization in skeletal muscle fibers labeled with rhodamine on the regulatory light chain. Biophys J 68:85S–86S

Anson M, Geeves MA, Kurzawa SE, Manstein D 1996 Myosin motors with artificial lever arms. EMBO J 15:6069–6074

Cooke R 1986 The mechanism of muscle contraction. CRC Crit Rev Biochem 21:53–118

Engelhardt WA, Ljubimowa MN 1939 Myosine and adenosine triphosphate. Nature 144: 668–669

Fisher AJ, Smith CA, Thoden JB et al 1995 X-ray structures of the myosin motor domain of *Dictyostelium* discoideum complexed with MgADP.BeFx and MgADP.AlF4$^-$. Biochemistry 34:8960–8972

Geeves MA, Conibear PB 1995 The role of threestate docking of myosin S1 with actin in force generation. Biophys J 68:199S–201S

Gollub J, Cremo CR, Cooke R 1996 ADP release produces a rotation of the neck region of smooth myosin but not skeletal myosin. Nat Struct Biol 3:796–802

Hill AV 1922 The mechanism of muscular contraction. Physiol Rev 2:310–341

Holmes KC 1996 Muscle proteins—their actions and interactions. Curr Opin Struct Biol 6: 781–789

Holmes KC 1997 The swinging lever arm hypothesis of muscle contraction. Curr Biol 7: R112–R118

Holmes KC, Popp D, Gebhard W, Kabsch W 1990 Atomic model of the actin filament. Nature 347:44–49

Houdusse A, Cohen C 1996 Structure of the regulatory domain of scallop myosin at 2 Å resolution: implications for regulation. Structure 4:21–32

Huxley AF, Niedergerke RM 1954 Structural changes in muscle during contraction. Interference microscopy of living muscle fibres. Nature 173:971–973

Huxley AF, Simmons R 1971 Proposed mechanism of force generation in striated muscle. Nature 233:533–538

Huxley HE 1957 The double array of filaments in cross-striated muscle. Biophys Biochem Cytol 3:631–648

Huxley HE 1958 The contraction of muscle. Sci Am 199:66–82

Huxley HE 1969 The mechanism of muscular contraction. Science 164:1356–1366

Huxley HE, Hanson J 1954 The cross-striations of muscle during contraction and stretch and their structural interpretation. Nature 173:973–976

Huxley HE, Simmons RM, Faruqi AR, Kress M, Bordas J, Koch MHJ 1981 Millisecond time-resolved changes in x-ray reflections from contracting muscle during rapid mechanical transients recorded using synchrotron radiation. Proc Natl Acad Sci USA 78:2297–2301

Irving M, Lombardi V, Piazzesi G, Ferenczi MA 1992 Myosin head movements are synchronous with the elementary force-generating process in muscle. Nature 357:156–158

Irving M, St Claire T, Allen C et al 1995 Tilting of the light-chain region of myosin during step length changes and active force generation in skeletal muscle. Nature 375:688–691

Jontes JD, Wilson-Kubalek EM, Milligan RA 1995 The brush border myosin-1 tail swings through a 32° arc upon ADP release. Nature 378:751–753

Kabsch W, Mannherz HG, Suck D, Pai EF, Holmes KC 1990 Atomic structure of the actin:DNase I complex. Nature 347:37–44

Kinose F, Wang SX, Kidambi US, Moncman CL, Winkelmann DA 1996 Glycine 699 is pivotal for the motor activity of skeletal muscle myosin. J Cell Biol 134:895–909

Kühne W 1864 Untersuchungen über das Protoplasma und die Contractilität. Verlag von Wilhelm Engelmann, Leipzig

Kurzawa SE, Manstein DJ, Geeves MA 1997 *Dictyostelium discoideum* myosin II: characterization of functional myosin motor fragments. Biochemistry 36:317–323

Lohmann K 1931 Darstellung der Adenylpyrophosphatsäure aus Muskulatur. Biochemistry 233:460–469

Lorenz M, Popp D, Holmes KC 1993 Refinement of the F-actin model against x-ray fiber diffraction data by the use of a directed mutation algorithm. J Mol Biol 234:826–836

Lymn RW, Taylor EW 1971 Mechanism of adenosine triphosphate hydrolysis of actomyosin. Biochemistry 10:4617–4624

Margossian SS, Lowey S 1973a Substructure of the myosin molecule: III. Preparation of single-headed derivatives of myosin. J Mol Biol 74:301–311

Margossian SS, Lowey S 1973b Substructure of the myosin molecule: IV. Interactions of myosin and its subfragments with adenosine triphosphate and actin. J Mol Biol 74:313–330

Milligan RA, Flicker PF 1987 Structural relationships of actin, myosin and tropomyosin revealed by cryo-electron microscopy. J Cell Biol 105:29–39

Moore PB, Huxley HE, DeRosier DJ 1970 Three-dimensional reconstruction of F-actin, thin filaments and decorated thin filaments. J Mol Biol 50:279–295

Nicholls A, Sharp KA, Honig B 1991 Protein folding and association: insights from the interfacial and thermodynamic properties of hydrocarbons. Proteins 11:281–296

Patterson B, Spudich JA 1996 Cold-sensitive mutations of *Dictyostelium* myosin heavy-chain highlight functional domains of the myosin motor. Genetics 143:801–810

Pollard TD, Bhandari D, Maupin P, Wachstock D, Weeds A, Zol H 1993 Direct visualization by electron microscopy of the weakly-bound intermediates in the actomyosin ATPase cycle. Biophys J 64:454–471

Poole KJV, Evans J, Rosenbaum G, Lorenz M, Cremo CR 1998 Micromolar ADP causes large-scale conformational changes in smooth-muscle myosin-S1 bound to actin, in prep

Rayment I, Holden HM, Whittaker M et al 1993a Structure of the actin–myosin complex and its implications for muscle contraction. Science 261:58–65

Rayment I, Rypniewski WR, Schmidt-Base K et al 1993b Three-dimensional structure of myosin subfragment-1: a molecular motor. Science 261:50–58
Reedy MK, Holmes KC, Tregear RT 1965 Induced changes in the orientation of the cross-bridges of glycerinated insect flight muscle. Nature 207:1276–1280
Schlichting I, Becker A, Manstein D, Holmes KC 1998 The structure of *Dictyostelium* myosin truncated at position 754 with bound ADP.BeF3, in prep
Schroeder RR, Manstein DJ, Jahn W et al 1993 Three-dimensional atomic model of F-actin decorated with *Dictyostelium* myosin S1. Nature 364:171–174
Schroeder R, Angert I, Jahn W, Holmes KC 1998 The structure of the actin–myosin complex determined by energy-filtered cryo–electron microscopy, in prep
Schweins T, Geyer M, Scheffzek K, Warschel A, Kalbitzer HR, Wittinghofer A 1995 Substrate-assisted catalysis as a mechanism for GTP hydrolysis of p21ras and other GTP-binding proteins. Struct Biol 2:36–43
Smith CA, Rayment I 1996a Active site comparisons highlight structural similarities between myosin and other P-loop proteins. Biophys J 70:1590–1602
Smith CA, Rayment I 1996b X-ray structure of the magnesium(ii).ADP.vanadate complex of the *Dictyostelium-discoideum* myosin motor domain to 1.9 Å resolution. Biochemistry 35:5404–5417
Straub FB 1943 Actin studies. Inst Med Chem Univ Szeged vol 2:3 (reprinted by S Karger, Basle–New York)
Szent-Gyorgyi AG 1953 Meromyosins the subunits of myosin. Arch Biochem Biophys 42:305
Uyeda T, Abramson PD, Spudich JA 1996 The neck region of the myosin motor domain acts as a lever arm to generate movement. Proc Natl Acad Sci USA 93:4459–4464
Whittaker M, WilsonKubalek EM, Smith JE, Faust L, Milligan RA, Sweeney HL 1995 Smooth muscle myosin moves 35 Å upon ADP release. Nature 378:748–751
Yount RG, Lawson D, Rayment I 1995 Is myosin a back door enzyme? Biophys J 68:47S–49S

DISCUSSION

Wolpert: How much more do you want to know? It seems to me that you have already solved the problem.

Holmes: I think we have the framework for understanding the problem. I would guess we need another 10 structures.

Wolpert: Why do you care?

Holmes: Because there is a large class of physicists and chemists who would like to know in detail where the energy is stored.

Rose: I think that muscle contraction is a marvellous example of the problems we are facing. What you've shown exquisitely demonstrates both the strengths and the limitations of the reductionist programme. You have shown how one can translate the physiology of the muscle twitch into the languages of biochemistry and physical chemistry, so we now know how a muscle twitches or contracts. However, there is another question that we are concerned about — the 'why?' question. This has to relate to the function of the muscle in its system as part of an organism. For instance, we could ask 'why is the frog muscle twitching?' Because it is jumping. 'Why is it jumping?' To escape from a predator. This is the complexity that we're dealing with here. Reductionism gives us one extremely important part of the answer to the questions that concern us as biologists, but we must concede that there are other questions which also constitute legitimate explanations of the phenomenon of the muscle twitch that we might want to try to answer.

Holmes: We've cut away everything, of course. We are just talking about the basic motor. We know it has an accelerator pedal, and it has brakes.

Rose: It's also part of a system.

Holmes: I've left all that out. We are going bottom up.

Wolpert: How far do you want to go? Do you want to explain muscles in terms of fundamental chemistry?

Holmes: My aim is to get this down to understanding how the energy of ATP hydrolysis gets turned into motion.

Ashmore: There is potentially another direction in which studies of muscle structure could go, which is to try to think about ways in which one doesn't end up with the same molecule — rather you might wish to end up with motors that go faster or slower, for instance. The image of myosin as a machine is very seductive. But are there ways in which one can step outside biology and think of ways of manipulating this molecule to do something completely different?

Williams: One problem with all machines is their limited efficiency. What is the efficiency of this machine, and why is stuck at a given efficiency?

Maynard Smith: Does anyone have any feel for the limits of thermal efficiency of motion of organisms? I seem to remember that it is rather low, about 25–30%.

Holmes: You can go up to 40–45% under ideal situations. I feel that's very high. A reversible engine can clearly be more efficient but tends to be slower. Muscle is actually very fast and throws away energy even faster: half the energy goes in phosphate release, and this makes the process unidirectional.

Garcia-Bellido: There are two components in understanding. One is what is called the synchronic component: for example, working out a particular muscle of the vertebrate and you know the nuts and bolts. The other is the diachronic one, where you trace back a phenomenon in evolution to more simple forms, where you find an invariant component by comparison, not by direct analysis of single pieces at a given moment in time. Have you got a diachronic understanding of muscle action — a general theory of how muscles work?

Holmes: A full answer to that question would take another talk. There is a recent paper by Cope et al (1996) where they compare 60 sequences, and these range from sunflowers through yeast to human. The core myosin is highly conserved, and I'm fairly sure the mechanisms are the same. I've only talked about part of the problem, namely, what happens on phosphate release, but we've not talked about what happens on ADP release. There is a whole linkage there which is more important for the processive use of myosin (i.e. not letting go of actin), which of course is the biggest use of myosin, for shunting vesicles around and so on. The primary mechanisms will probably be the same, but these processive myosins have a second mechanism which controls ADP release that we don't understand yet.

Dover: The diachronic approach is more than just saying that there is a given ancient function which has been conserved. In the case of actin, for example, it is about trying to figure out what happened to actin throughout its evolutionary history. Actin and myosin must to some degree have co-evolved, for the unit of evolution is the actin–myosin interaction. In other words there must have been a history of events which reflect the functional requirements for this interaction in the whole organism which will influence exactly how actin and myosin work. Do you have any feel for how all this got to be the way it is?

Holmes: Rather simplistically I assume that all myosins are basically alike, and then there are decorations on these that depend on the function. The primary function seems to be in cytokinesis, and since all organisms require this it is an old function. But then of course myosin is useful in cell biology: new myosins are being turned up by cell biologists at the rate of two or three a week. They are nearly always associated with actin: I don't think they would have any function without actin. Thus actin and myosin co-evolved. The core myosin is highly conserved, but actin is even more highly conserved. The variation between yeast and human actin is only 20 amino acids in 350. Actin binds to at least 20 proteins, so it has co-evolved with a large number of proteins and it has not had the chance to alter.

Garcia-Bellido: I didn't want to know the variation: rather, I was concerned with knowing which are the dispensable parts of the structure, which is the core of the variance. It is crucial to know what the structure is without decoration. This is the meaningful bit, and this is the power of the diachronic comparison. Thus my question is, in the primitive forms, do you still see the same process of the movement of the myosin over the actin?

Holmes: Yes.

Quinn: Is it likely that these tricks are used by other motility molecules such as kinesin?

Holmes: Kinesin is very similar to myosin, but so far we only have one crystal structure. The assumption is that it is also like a G protein. It doesn't have a tail. The primary mechanism is going to be similar but what is done with the energy at the back end is different.

Dover: There is a parallel phenomenon in biology to this actin–myosin story. Proteins were discovered which carry what are called homeodomains which bind to target sites in DNA and induce gene expression. When they were first discovered they were hailed as the Rosetta Stone of developmental biology. We now know, of course, that there are many other different motifs in proteins which can bind to DNA. But even if you just limit the discussion to the homeodomain itself, this can be found as a little block of amino acid sequences in many totally unrelated proteins and in all sorts of organisms, showing that is an ancient conserved function. So, detailed knowledge of the biophysics of homeodomain–DNA binding (and it's important we know that) tells us nothing at all about why all these different genes carry the homeodomain, why they bind to particular target sites in front of other unrelated genes, and why all this unfolding of gene interactions occurs, for instance, in the development of *Drosophila* or *Caenorhabditis*. There is a level at which biophysics tells us nothing about why there is a subset of proteins carrying homeodomain motifs and another subset carrying other DNA binding motifs, or why some genes carry homeodomain target sites in their regulatory regions and other genes carry target sites to other DNA binding motifs.

Wolpert: That is quite a different class of question: it is an evolutionary question. Some people think that you can't separate evolutionary and mechanistic questions. But you can: there wasn't a word about evolution in Ken Holmes' talk. He was looking at the mechanism of muscle contraction.

Maynard Smith: I agree, but I disagree with the notion that one of those questions can be given a reductionist answer, and the other one can't.

Reference

Cope TV, Whisstock J, Rayment I, Kendrick-Jones J 1996 Conservation within the myosin motor domain: implications for structure and function. Structure 4:969–987

Reductionism and explanation in cell biology

Paul Nurse

Imperial Cancer Research Fund, 44 Lincoln's Inn Fields, London WC2A 3PX, UK

Abstract. It is likely to be impossible or very difficult to provide a detailed description of the molecular interactions underlying all cellular phenomena. However, methods and ways of thinking are now available or being developed that can deal better with the complexity and greater extension in space and time found at the level of the cell. This will lead to the identification of some components or groups of components as being of particular importance for a cellular phenomenon which can then be studied in detailed molecular terms. In other cases detailed molecular characterization may be replaced by a logical description of the process which emphasizes the information flow and processing rather than the nature of the individual components and their interactions. This may provide an adequate explanation for an appropriate understanding of the cellular phenomena involved.

1998 The limits of reductionism in biology. Wiley, Chichester (Novartis Foundation Symposium 213) p 93–105

Reductionism is the principle of analysing complex things into simpler more basic constituents. In biology this has led to the view that living processes can be explained in terms of the material composition and physicochemical activities of living things. This approach has been enormously successful and many of the triumphs of modern biology are a consequence of using methodologies that treat living characteristics as if they can be adequately explained by reductionism (Bullock & Stallybrass 1977). In my opinion this view cannot be challenged in epistemological terms, but the 'as if' qualification is of crucial importance. Although it has been argued in principle that it may be possible to explain living things completely in terms of their basic components, this may neither be practically possible nor even necessary to provide an adequate explanation of living phenomena. It is these two assertions which I want to explore in this essay, primarily in the context of cell biology.

Life, molecules and cells

Cells are of particular importance when discussing reductionism in biology because the cell is the most basic level at which the major characteristics of life emerge. The simplest autonomous living organisms are single-celled and yet exhibit all the properties of life. They are capable of multiplication, variation and heredity, and thus can undergo evolution by natural selection (Muller 1966). As a consequence they can acquire the adaptations and functions necessary for self organization and reproduction characteristic of living things (Maynard Smith & Szathmáry 1995). Cells can survive autonomously in a simple environment through development of a sophisticated metabolism which organizes and maintains them as independent entities. An important goal for cell biologists is to provide an adequate explanation of how a cell achieves this organization. The basic constituents of cells are molecules, and the usual methodological approach employed by cell biologists has been to attempt to describe the phenomena which emerge with cellular organization in terms of the molecular interactions involved. Studies of molecular interactions *in vitro* generally deal with a limited number of different molecular types interacting with each other at a short range in the order of the length of a chemical bond, and over a limited time span. In contrast, the scale of a cell extends in space, time and complexity far beyond that normally observed with these simple *in vitro* molecular interactions. Understanding these extensions is important for appreciating the difficulties which arise when trying to interpret cellular organization in terms of a complete description of molecular interactions.

Macromolecular assemblies and machines

Extension on a modest scale is found with macromolecular assemblies and machines such as the ribosome carrying out protein synthesis, the replisome carrying out DNA synthesis, the macromolecular assembly which forms a phage, and signalling pathways communicating signals from the cell surface to other components within the cell. The different macromolecules making up these assemblies and machines can be studied *in vitro* as single components or as subsets of components. This yields useful information on the types of reaction that can occur and on their kinetics, but this information can be misleading because the conditions found within the macromolecular assembly may not be the same as those which apply to the operation of a few components in dilute solution, which are the conditions used in classical biochemical *in vitro* analysis. A number of differences can be identified. Firstly, the assembly of macromolecules into complexes allows the channelling of intermediates to occur. This has been described in the context of intermediate metabolism when a low molecular weight component undergoes several sequential chemical changes within a complex, and is also of significance when dealing with macromolecular reactions. Channelling allows intermediates to be passed on from one component to another without free exchange with the region surrounding the complex, and this allows reactions to occur which would not be energetically favourable in free solution. Secondly, the conditions present within a macromolecular complex may differ substantially from those generally found *in vitro*. The concentration of protein may be up to a hundred times greater leading to a significant exclusion of water, and ionic conditions may also differ. Such differences will certainly influence the kinetics of reactions and possibly also the nature of reactions that can take place. Thirdly, the existence of a complex results in the isolation of different parts of the system from each other. This allows intermediates in different parts of the complex to behave independently of each other, which may be necessary to achieve a certain set of reactions, and may be difficult to accomplish in free solution.

These differences mean that there will always be limitations to conventional *in vitro* biochemical analysis when applied to macromolecular assemblies and machines. Study of the individual components or subsets of components is important but may miss essential characteristics of the complete system. The local organized microenvironment changes molecular conditions (Welch 1989), and it is necessary to understand how the structure in both space and time within cells influences the molecules and processes involved (Rose 1988).

Extension in space and time

More problems arise when greater extension is considered at the level of an organelle or a whole cell where scale extends far beyond that observed during molecular interactions. It is clear that compartmentation within different organelles or parts of the cell can influence cellular behaviour. At a greater scale is the organization of the entire cell in space which implies the existence of morphogenetic mechanisms such as spatial fields and diffusion gradients which can impart positional information to the cell. Likewise, greater extension in time gives rise to new phenomena. It is becoming increasingly clear that the kinetics of signalling pathways may determine the nature of the signal sent. A useful analogy is telegraphic transmission of morse code where information is communicated by the duration and order of electrical pulses and as a consequence is a much richer vehicle of communication than would be the case if a simple on/off electrical signal was employed. Context in regulatory circuits is also of great importance. The same signalling process may lead to quite different outcomes depending on the particular cell under study suggesting that the context within which the signal is sent plays a crucial role (Lloyd et al 1997). Oscillatory behaviour has also been observed which can have regulatory implications (Goldbeter 1996) in intermediate metabolism, gene regulation and the cell cycle. At even greater extensions in time are found circadian rhythms and cellular memory as shown during differentiation.

None of these phenomena are well understood and yet they provide us with some of the most fascinating problems of cell biology. It is important to use and develop methods and ways of thinking that will allow analysis of these problems. In many situations the aim should be to provide explanations in terms of molecular interactions but this is not always possible, and explanations at a different level should then be sought. In the next sections of this essay, I want to consider how these aims may be achieved.

Biochemical approaches

The most straightforward approach for extending conventional *in vitro* assay systems is to work on purified or partially purified macromolecular assemblies or machines. One of the first examples of this approach was the establishment of an *in vitro* DNA replication system using purified proteins from *Escherichia coli* (Kornberg & Baker 1991). This relied on the addition of polyethylene glycol to reduce the level of water and thus concentrate the protein components, and also on the use of more physiological buffers. In the same tradition a reconstituted system has been recently described which can carry out nucleotide excision repair (Sijbers et al 1996). Another example is the development of *in vitro* protein translation systems which have not utilized reconstituted purified components, but have been based on relatively crude but concentrated cellular lysates to achieve activity.

This general approach has been employed a great deal to derive activities from lysates of *Xenopus* extracts which have many stored products. Once again a key step for success has been to maintain a high concentration of proteins in the lysates. Because eggs have stores of components required for many cell divisions, such extracts are a very rich source of cell cycle activities, and quite sophisticated cell cycle assays have been developed. For example, DNA is replicated only once per cell cycle and this is also the case in the crude lysates prepared from eggs which undergo DNA replication. Manipulation of these lysates has allowed a crude activity to be identified called 'licensing factor' which is necessary to allow a second round of DNA replication to occur (Chong et al 1995). Having identified an activity in crude lysates, further purification became possible which led to the characterization of a macromolecular complex responsible for the activity (Rowles et al 1996).

The use of crude lysates to establish complex assays reflecting whole cell processes provides a powerful means to investigate cellular phenomena, and in some cases can lead to the preparation of more purified macromolecular complexes. As is clear from the above example, it is important to select the appropriate cellular sources for these activities depending upon the phenomenon of interest. If it is impossible to recover an activity in lysates then an alternative is to use permeabilized cells which retain some aspects of total cellular structure. In this case the properties of the entire activity can be studied even though the behaviour of the individual components making up the activity may not be readily accessible.

Cellular approaches

The next level up in organization is to study a cellular phenomenon in an intact cell. Usually this requires the development of techniques that work on limited numbers of cells and as a consequence generally utilizes microscopic procedures. Cell injection of a component of interest followed by microscopic inspection of cellular behaviour is a well-established procedure. Two recent developments will enhance this type of approach further. The first is the use of naturally fluorescent protein tags such as the jellyfish green fluorescent protein (GFP). The gene encoding the GFP tag can be fused to any gene of interest and expressed in cells. This allows the monitoring of the position of the GFP–protein fusion within cells and allows location to be studied in real time. Although there are drawbacks to the use of a tag of 25 kDa and only a rather approximate position can be monitored, this procedure represents a major advance for the study of processes extended in both space and time within the cell. The second development is fluorescence resonance energy transfer (FRET) microscopy (Bastiaens & Jovin 1996). With this method, energy is transferred from an excited donor fluorophore to an acceptor chromophore which then emits light at a wavelength differing from the initial excitation wavelength. Transfer is only possible if donor and acceptor are within 5–10 nm, and so if the fluorophore and chromophore are located in different proteins then FRET can be used to assess if the two proteins are located within 10 nm in the cell. This technique holds much promise for the future. Once it has been developed fully for use in living cells, it will be possible to follow in real time and space when and where two proteins closely interact. After such an interaction a light signal is emitted allowing the dynamics and position of the interactions to be monitored simply by following the 'sparks' in the cell. An obvious application is during signal transduction when the dynamics of signalling can be assessed which, as explained earlier, can contain much information missed by most other methods.

Genetic approaches

Although genetics is often criticized for putting too much emphasis on a reductionist approach, it actually provides one of the most powerful methodologies for the study of complex phenomena within cells. Procedures are available which allow identification of the components which make up regulatory networks. These include yeast two-hybrid analysis to identify gene products which directly interact with each other physically (Fields & Song 1989), and genetic interaction analyses such as extragenic suppression and synthetic lethalities. Such procedures allow the identification of components making up a network, but too often investigators stop at this early stage of 'naming the parts' when really it needs to be followed by extensive functional analyses to give a satisfactory explanation of the mechanisms involved.

Genetics also provides a very precise means to perturb or alter processes within the whole cell by specific mutation of particular genes which generate precisely altered proteins. With this approach only one element is initially perturbed but the effects of this local perturbation on global behaviour can be examined by a the combination of genetic mutation with physiological investigation. A very good example has been the elucidation of the controls acting over the cell cycle. This work started with the identification of cell cycle mutants which were then studied physiologically at the level of the intact cell to determine exactly how they affected cell cycle progression (Nurse 1990). Next the biochemical activities associated with the relevant gene products were identified by a combination of sequence analysis and conventional biochemical assays, which allowed the molecular functions underlying cell cycle control to be described. Similar approaches have been used to study many other cellular phenomena including protein transport, cell morphogenesis and circadian rhythms.

Genetic methods are also potentially very useful for connecting the *in vitro* assays of conventional biochemistry to higher levels of organization. For example a mutated protein can be studied *in vitro* and shown to have altered kinetic behaviour, such as a changed V_{max} or K_m. The relevance of such changes can then be investigated by studying the effect of this mutated protein on the overall activity of a purified macromolecular complex or cell lysate, or in the intact cell itself. The point once again is that overall global behaviour of a system or subsystem is assessed following a very specific change of a particular component.

Conceptual approaches

It will be an immense if not impossible task to adequately describe cellular phenomena in terms of a precise description of all the molecular interactions involved. Already there is an information overload in cellular and molecular biology with many molecules identified but with the underlying processes much less well understood. In this section I want to consider possible conceptual approaches providing alternative levels or types of description which may produce adequate explanations of cellular phenomena without a full molecular characterization.

Firstly, it is necessary to emphasize that not all components within a cell are of equal interest with respect to a particular cellular phenomenon. The argument here owes much to the study of flux through a metabolic pathway. It has been shown that although the control of flux through a pathway is distributed amongst all the steps of a pathway, some of the steps play a much more significant role in determining the rate of flux than do others. This is revealed by the following thought experiment whereby each step is perturbed by a small amount, say up or down by 10%, and the effect on the overall flux through the pathway is assessed. The major rate-limiting or pacemaker steps can be revealed by such analysis. Thus if control of flux is the phenomenon of interest then attention should be focused on the major rate-limiting steps. Applying this approach to the control of the cell cycle resulted in the identification of those steps which were rate limiting for cell cycle progression, rather than those that were simply required for cell cycle progression (Nurse 1975). The concept being developed here is that of probing the phenomenon of interest to identify those elements which are of greatest importance, rather than to try to describe every element that is involved. This requires study of the system as a whole but then moves on to an in-depth investigation of only a limited set of key components.

Another approach is to use modelling. Two types of procedure are possible. In the first, kinetic parameters are assigned to each component of a process on the basis of experimental measurement, usually obtained by conventional *in vitro* biochemical assays. These are then modelled and the subsequent differential equations solved to produce overall behaviour of the system. This bottom-up approach has been useful in thinking about cell cycle control, for example (Novak & Tyson 1995), but does rely on having complete and accurate data concerning the kinetic behaviour of the individual components. Despite this limitation it is useful for clarifying thoughts about problems and can be very helpful if unexpected behaviour is predicted which can then be tested experimentally.

The second procedure has yet to be used much but may have greater potential in the future. This is to represent the different steps in a process as information tools analogous to components in an electrical circuit. The idea is not to precisely model kinetic parameters, but to identify each element as an information processing component and to establish which elements are linked and in what way. Each component will have a property or collection of properties such as amplification, integration, timing, information storage and feedback (Bray 1995). Sufficient information may be provided by the sequence of the gene encoding a component to have some idea about what information processing is possible, and knowledge about the overall behaviour of the processes may also limit the possibilities open to individual components. For example, G proteins can exist in two states depending on whether GTP or GDP is bound, and the transition from one state to another is tightly regulated. Such a device can act as a timer with a proof-reading function in a copying process such as protein translation (Kirkwood et al 1986), or can act as an amplifier in a signal transduction pathway if it generates a signal as long as it is in one state or another. Modelling in this way might profit from work with artificial intelligence and neural networks.

Application of control theory to the analysis of regulatory networks would emphasize the need to have a control system which is robust in operation. Robust behaviour allows perturbations of particular components or shifts in behaviour beyond normal parameters to be accommodated without the system failing. However, it is important to remember that cellular processes have been developed by natural selection and so often will include subsystems which are 'add-ons' to pre-existing subsystems, giving rise to redundancy. It is becoming clear that there is extensive redundancy in biological systems which is not normally the case in man-made control systems, except for simple systems duplication to deal with individual component breakdown. The phenomenon of redundancy must be fully recognized in analyses of this sort and in subsequent experimental investigation. A further development would be to focus on cellular phenomena as 'black boxes' with specific output behaviour as a consequence of specific input behaviour. With this approach there may be no need for a detailed description of what makes up each individual black box, but only of the overall behaviour of each black box. This can be determined by the appropriate monitoring of the input and output signals working with the intact cell or macromolecular machine derived from a cell. Ignoring the nature of the 'substances' in the black boxes may emphasize too much the view of the cell as composed of information processing devices, but may provide an adequate explanation of phenomena in certain circumstances.

Also of interest is the possibility that complex behaviour can emerge from the operation of a network controlled by simple rules. There has been much study of this by theoretical biologists (Kauffman 1993), although the tendency has been to use switching networks more as a metaphor for emergent behaviour rather than attempt a realistic logical modelling of control circuits. It has been concluded from these studies that the previous history of a regulatory network can play a role in establishing present behaviour. This is relevant for the phenomenon of context mentioned earlier when discussing signalling. The consequence of a particular input signal will be dependent on the state of the network, which in turn is dependent on its previous behaviour.

It has also been proposed that different sorts of questions should be considered when trying to explain biological phenomena. One example is the development of more general ensemble theories which define the generic properties of regulatory circuits (Kauffman 1993). The notion here has been that certain regulatory states are more stable and as a consequence biological systems will tend to occupy only a limited set of attractor states. If the nature of these states and the reasons why they exist could be understood then this may be helpful in explaining cellular phenomena. The general point to be emphasized is that temporal and spatial order in biological systems and cells in particular may give rise to only a limited set of stable possibilities. If the restraints underlying these limitations are better understood, then analyses do not have to consider all imaginable behaviours but only those which are stable and therefore more likely to occur. If such views are correct then the numbers and types of models that need to be considered can be restricted, simplifying the subsequent analysis of complex cellular phenomena.

Acknowledgements

I would like to thank Jacky Hayles, Hartmut Land and John Tooze for their helpful comments on this manuscript.

References

Bastiaens PIH, Jovin TM 1996 Fluorescence resonance energy transfer (FRET) microscopy. In: Celis JE (ed) Cell biology: a laboratory handbook, 2nd edn
Bray D 1995 Protein molecules as computational elements in living cells. Nature 376:307–312
Bullock A, Stallybrass O 1977 The Fontana dictionary of modern thought. Fontana & Collins, London
Chong JPJ, Mahbubani HM, Khoo CY, Blow JJ 1995 Purification of an MCM-containing complex as a component of the DNA replication licensing system. Nature 375:418–421
Fields S, Song O 1989 A novel genetic system to detect protein–protein interactions. Nature 340:245–246
Goldbeter A 1996 Biochemical oscillations and cellular rhythms. Cambridge University Press, Cambridge
Kauffman S 1993 The origins of order. Oxford University Press, Oxford
Kirkwood T, Rosenberger R, Galas D 1986 Accuracy in molecular processes. Chapman & Hall, London
Kornberg A, Baker T 1991 DNA replication. Freeman, New York
Lloyd AC, Obermuller F, Staddon S, Barth CF, McMahon ML 1997 Cooperating oncogenes converge to regulate cyclin/cdk complexes. Genes Dev 11:663–677
Maynard Smith J, Szathmáry E 1995 The major transitions in evolution. WH Freeman, Oxford
Muller H 1966 The gene material as the initiator and organising basis of life. Am Nat 100:493–517
Novak B, Tyson JJ 1995 Quantitative analysis of a molecular model of mitotic control in fission yeast. J Theor Biol 173:283–305
Nurse P 1975 Genetic control of cell size at cell division in yeast. Nature 256:547–551
Nurse P 1990 Universal control mechanism regulating onset of M-phase. Nature 344:503–508
Rose S 1988 Reflections on reductionism. Trends Biochem Sci 13:160–162
Rowles A, Chong JP, Brown L, Howell M, Evan GI, Blow JJ 1996 Interaction between the origin recognition complex and the replication licensing system in *Xenopus*. Cell 87:287–296
Sijbers AM, Delaat WL, Ariza RR et al 1996 Xeroderma pigmentosum group F caused by a defect in a structure-specific DNA repair endonuclease. Cell 86:811–822
Welch GR 1989 Of men, molecules, and (ir)reducibility. Bioessays 11:187–190

DISCUSSION

Hess: In my own studies of periodic reactions and periodic spatial propagation phenomena—especially in the case of the Zhabotinski reaction of glycolysis, the aggregation process of the slime mould *Dictyostelium discoideum*, as well as in the case of the myocardial pulse propagation process—it became obvious that complex macroscopic dynamics in biology and chemistry can be understood well by reducing and lumping together the many elementary chemical or biochemical reaction steps describing such processes in detail to yield two or three coupled differential (or partial differential) equations with two or three variables. Such global equation systems display the large variety of dynamic features, common to all non-linear systems, such as steady states, limit cycle oscillations, quasiperiodic and chaotic behaviour, synchronized coupling with amplitude and frequency modulation and, coupled with diffusion, spatial wave propagation in the form of spirals, toroids and others in two and three dimensions.

Here it is important to note that with increasing complexity of the biochemical networks involving allosteric enzyme reactions as well as complex interconversion, receptor binding, membrane transport and gene interaction processes, the time and space scale of the resulting phenomena is appreciably extended. Thus, the period of the oscillatory regimes at various organizational levels of biology covers the range between a few microseconds up to a circadian 24 h scale, and the space scale bridges the micron to the centimetre scale (Hess 1997).

Garcia-Bellido: One of the things that makes cell biology (and by extension, development) difficult, is that many of these processes take place in parallel, and we don't know their connectivity. This makes the problem extremely difficult computationally.

Nurse: It is difficult, but we mustn't be overwhelmed by the difficulty. If we can identify certain principles, we may be able to limit where we have to search for all the solutions. Both for evolutionary reasons and for these other 'what is actually stable in the real world' reasons, we may be able to limit the possible variations that we have to search for. This may make the task computationally more straightforward.

Bray: One of the interesting questions that came up in the discussion yesterday was the distinction between a description and an explanation of a phenomenon. I feel that our work throws some light on that distinction. Paul Nurse mentioned that living cells detect their environment and send signals to other cells through a series of biochemical circuits that are bewildering in their number and complexity. A human cell has tens of thousands of receptors on its surface, which may be of hundreds of different kinds. On the other side of the membrane they send signals to dozens of different G proteins and SH2/SH3 binding proteins, and there are multiple protein tyrosine kinases and phosphatases further downstream. Lewis

Wolpert yesterday quoted an estimate by Tony Hunter that 5% of the human genome codes for protein tyrosine kinases and that something like half of the genome may be involved in cell signalling. And it is not only the large number of components that is confusing, but also the fact that they don't work in the linear pathways but are all connected to each other. It is important to try to come to grips with this — it is a real crisis, in a way, because this is the nervous system of the cell, and if we are going to understand how cells move, differentiate and so on, we are going to have to understand this part of their anatomy.

The approach that I and my colleagues have taken during the last six years is to focus on a particular well-defined signal pathway, which is the one used in bacterial chemotaxis (reviewed in Parkinson 1993, Eisenbach 1996, Stock & Surette 1996). It is a small pathway built up of seven proteins, that performs a specific computation. The pathway detects the extracellular concentration of an attractant, transmits the signal, amplifies it and also differentiates it, and then sends the resulting signal to the flagella motors which control the swimming of the bacteria. All of these seven proteins have been sequenced and the structures of two of them are known with more on the way. The kinetic parameters of many individual proteins have been characterized *in vitro*. Perhaps most important of all, there's a large list of mutants in which these signalling proteins are either knocked out or up-regulated singly or in combination. The phenotypes of these bacteria have been identified, against which we can test any model we build.

We have launched into a detailed computer-based model of this system, using the available information and trying to match it to the experimental data. My colleague Bob Bourret has set up experiments to test predictions coming out of the model. We are of course interested in proving that the model is correct and in making testable predictions. However, I have to say that the underlying drive of this exercise is to search for ways in which we can come to grips with this immense complexity of cell signalling processes in general.

I will just mention one result of this study that illustrates how it can help us analyse and ultimately understand the chemotactic pathway (Bray & Bourret 1995). In the course of developing our computer simulations, we came across a number of mutants which had the 'wrong' phenotype — that is, the computer model disagreed with the experiment. We realized that this was because a number of the proteins, perhaps five of the seven, are part of a complex associated with the membrane. It wasn't until we built into our model the series of steps leading to the formation of this complex that we were able to obtain the correct phenotype of these puzzling mutants. This drove home to us the fact that although this biochemical pathway involves seven proteins, they aren't simply floating around free in aqueous solution. In fact, there's a lot of structure there. Indeed, it is possible to think of bacterial chemotaxis as involving two large protein complexes. One is the flagella motor, the other is the receptor complex, and communication between these two depends on just one freely-diffusing protein. The signalling complex is interesting and sophisticated: it detects the extracellular concentrations of aspartate and maltose which are attractants; it detects the extracellular concentrations of nickel which is a repellent; it is also involved in the detection of hydrogen ion concentration and pH chemotaxis; and it is also claimed to be involved in thermo detection. Now these environmental stimuli are all imposed on the outside of the complex. On the cytoplasmic side, the complex monitors the recent past of the bacterium by being methylated by cytosolic enzymes. The whole gemisch is integrated and the resultant is the level of phosphorylation which affects the concentration of the diffusible element and hence the flagella motor.

Bacteria are not alone in this respect: signalling complexes are abundant in eukaryotic cell signalling pathways. They are usually associated with the membrane or cytoskeleton, and sometimes they're only formed transiently. But I think that they provide us with an intermediate level of explanation between signalling molecules and entire pathways. Signalling complexes are small solid-state computational elements which work quickly and efficiently. They don't have to rely on diffusion, which is a tremendous advantage, and they are also specialized for specific functions. They send their signals by generating diffusing molecules which can spread around to other targets in the cell.

Returning to the issue of reductionism, it seems to me that signalling complexes could eventually help us understand even highly complicated pathways of intercellular signals. As illustrated in the case of bacterial chemotaxis, signalling complexes can reduce a long list of protein species catalysing multiple binding and catalytic reactions to a single pathway composed of a small number of processing units and signalling species. In a sense, you could say that they take us from a 'description' to an 'explanation'—the distinction between the two lying mainly in the number of elements. Only if the number of elements becomes sufficiently small can we hope to manipulate them mentally and hence achieve an intuitive understanding.

Bateson: You have described this in terms of a stimulus–response mechanism. Is there any negative feedback? Why don't they go on beating their flagella?

Bray: The system shows adaptation: this is the methylation I briefly described. The stimulus response is very fast (a few tenths of a second) but that sets in motion a slower response which takes place over a minute or so in which the receptor is slowly modified on the cytoplasmic side by methylation. This chemical modification adapts the response and so brings it back to the steady state. Bacterial chemotaxis is remarkable, because it can adapt over five orders of magnitude of concentration.

References

Bray D, Bourret RB 1995 Computer analysis of binding reactions leading to a transmembrane receptor-linked multiprotein complex involved in bacterial chemotaxis. Mol Biol Cell 6:1367–1380
Eisenbach M 1996 Control of bacterial chemotaxis. Mol Microbiol 20:903–910
Hess B 1997 Periodic pattern in biochemical reactions. Q Rev Biophys 30:121–176
Parkinson JS 1993 Signal transduction schemes of bacteria. Cell 73:857–871
Stock JB, Surette MG 1996 Chemotaxis. In: Neidhardt FC (ed) *Escherichia* and *Salmonella*: cellular and molecular biology. American Society for Microbiology, Washington DC, p 1103–1129

Biological computation

Sydney Brenner

Molecular Sciences Research Institute Inc., 9894 Genesee Avenue, La Jolla, CA 92037, USA

Abstract. It is argued that biological systems can be viewed as special computing devices. This view emerges from considerations of how information is stored in and retrieved from the genes. Genes can only specify the properties of the proteins they code for, and any integrative properties of the system must be 'computed' by their interactions. This provides a framework for analysis by simulation and sets practical bounds on what can be achieved by reductionist models.

1998 The limits of reductionism in biology. Wiley, Chichester (Novartis Foundation Symposium 213) p 106–116

My title is deliberately ambiguous but the two meanings are intimately connected. One is that living organisms can be viewed as special computing mechanisms. We first need to gain an understanding of what might be called their 'computational style' and then we can move on to the second aspect which might be better called 'computational biology'. This involves the design of appropriate computational models that simulate the behaviour of particular biological systems, and which provide a theoretical explanation of them.

Essentially this is a form of pragmatic reductionism. We propose to reduce our problems only to the level from which useful computations may be made and we separate this from the general reductionist programme which aims to explain everything in terms of general physical principles. Although I am convinced that Schrödinger's equation underlies the behaviour of everything, it does not provide a useful basis for the construction of motor cars or bridges. On the other hand, it is certainly relevant to compute the properties of the hydrogen atom from it. What we are looking for in science is explanation and understanding, which provide us with the ability to make predictions, that is, calculate outcomes which go beyond what we have directly observed. Thus, once I have a model, I can test its validity by changing it in some way and then look for the predicted outcome by making the same change in the real system.

In what sense may we look on living organisms as special computers? First, I want to make a distinction between the superficial use of what might be called the 'computer metaphor' and a much narrower technical description. There is a view that the whole of nature is one great computer: computation abounds everywhere in the universe, and the objects we study have, so to speak, captured bits of this computational power and used it. In a sense this is what all mathematicians believe, but I find that this is, like all other statements of great generality, absolutely true but totally vacuous. It does not help me with my problem, which is how to make a mouse, for example. If I wish to go beyond the metaphor, I have to justify why I have chosen computation as my mode of explanation and why I am departing from the more familiar form based on physical mechanisms. The reason stems from a basic property shared by all living systems. These are totally unlike all other natural complex systems, in that they carry an internal description of themselves written in their genes. It is this description which is passed on from generation to generation and from which the organism is 'computed'. If we compare this to the weather, for example, we find that there is no internal description of the weather that we can separate physically from the weather itself. For the weather we need the physics of matter and energy, but the existence of DNA implies something new; it is the physics of information, that is, computation.

It was a failure to appreciate this difference which misled Schrödinger in his book *What is Life?* He was clear that the genetic material contained a programme for the development of the organism, but he thought that the genes also contained the means for its execution. They do not contain the means, but, rather, a *description* of the means for execution. This was precisely the distinction made by John von Neumann in his theory of self-reproducing machines, and shown to be a necessary feature of such automata. The means to translate the instruction tape is obtained from the parent machine and is used to read the description of the means and so install the means in the daughter machine. In biological systems, the egg has the means to read the genes, and the new organism makes new eggs. Thus, in addition to DNA, there is a physical continuity of the reading machinery over the total course of biological evolution, but the informational continuity is preserved in the genes.

In order to specify more narrowly what I mean by a biological computer, I want to give one example which will also illustrate what may be called the computational style of a biological system (see Brenner 1981). Consider a bacterium, such as *Escherichia coli*, growing in a simple medium. Every 30 minutes or so, each bacterium must make another one, synthesizing everything from simple components. We can consider it as a chemical factory with a few thousand different chemical reactions going on in parallel and all mixed up in the same space. How is everything regulated so that coherence is maintained in the face of what appears to be an enormous complexity? Now if this were a man-made factory, a control engineer would do something along the following lines. First, the plant will have been designed with pipes carrying intermediates from one reactor to the next. There would be a 'wiring diagram' and each reactor would have a physical address. Then there would be a large number of sensing probes which would continuously measure rates of reaction at each particular step, the rates of flow along the pipes, and perhaps other parameters such as temperature and pressure. Information from these probes would be carried back to a central computer which would calculate a global optimum and, in order to effect any necessary changes to meet that optimum, would send back signals to a set of valves which could alter flow rates, or other effectors which might change the temperature or pressure. Since changes in one segment might affect processes in other segments, the computer would need to make continuous adjustments. It is known that when such systems reach some level of complexity, it becomes difficult to maintain them in a steady state: they become metastable, and either run away and blow up, or lie down and die. The main problems are the enormous amount of computation required to maintain the system and the cumulation of inaccuracies. Even if the central machine is assisted by lower level computers that will translate the sensors or control the valves, it will have to supervise them as well as administer itself and this alone consumes a large sector of the computational resource. It is also clear that in order to write the program that runs the system, a programmer would have to understand it in some detail; the program would therefore include the theory of the system. However, as we all know, it is extremely difficult for us to understand systems where many different things are going on at the same time. It would be difficult to write and comprehend a program with (while, X,(while, Y,(while, Z, . . . embedded more than three levels deep. Our brains deal with only few things at the same time, and can handle serial processes but not parallel ones. Indeed, mathematics could be thought of as a special language designed to deal with these problems and when systems are linear, with entities that can be treated statistically as in parts of physics, mathematics provides the theoretical framework from which predictions may be made and tested. There are mathematical tools for optimization problems, but our factory will have many nonlinear functions in it.

Let us now see how the bacterium handles the problem. Firstly, there is no direct characterization of the global properties of the systems, that is, the interactions are not explicitly defined. There are no physical addresses, and everything proceeds in homogeneous solution. In the small volume of a bacterium, diffusion is fast enough to put all small molecules everywhere; the transit time for a molecule of molecular weight 500 is about 1 ms. The reactors are enzyme molecules and the elementary events are molecular collisions. Even though most of the molecules colliding with an enzyme are the wrong ones and even though most of the collisions of the right one are in the wrong place, the number of random collisions is so large that a given substrate will find its binding site on the enzyme with reasonable frequency. The bacterium does not direct products to reactions through fixed wires or pipes, but through the specificity of the binding site, implementing a *logical address* so that the broadcast system can be effectively used. When, in a conversation with the mathematician, Kurt Gödel, in Princeton in 1972, I pointed out the nature and the importance of the binding site, he said: 'This is the end of vitalism'.

The binding site is also the means whereby control systems are implemented in a bacterium. The processes are grouped into small sets or modules, and each pathway is controlled be feedback inhibition of the rate of the first enzyme of the pathway by the final product. There is no central computer, no master program and no supervisor. The system works because all the components do their job correctly and satisfy the global requirements.

I have left one question open. If this is the way it works, then to explain it I have to account for how it got that way. The factory has a designer, but the bacterium does not, and in this approach, functional explanations must be accompanied by evolutionary explanations. Indeed, we have already had to do this in accounting for the separation of the description of the means of reading the genes from the means themselves.

We now want to define the general nature of biological computers. Discard all the analogies that have been discussed before and do not be influenced by the computers that we use today. Imagine now that we have two machines which we will call P and T, and which provide us with answers to questions involving mathematical functions. We can ask each for the values of squares, square roots, logarithms and so on, and so we can type into each machine, for example, the query, factorial(5). Both machines will reply, 120. Now we look inside each computer. In P we find many encoded programs, like most present day computers. Thus the function, factorial(X), might be recursively defined: (multiply, X, (if X = 1, then X, else (factorial(X-1)))). If we look at the system programs of P we find that factorial(X) is translated into: Invoke the program labelled factorial with parameter 5. T is completely different. It has a huge store of tables. There are tables of logarithms, square roots and factorials much like the tables we used before the age of calculators, computers and slide rules. The system program of T translates factorial(5) into: Look up the table labelled factorial at position 5, where the answer 120 is stored. T is a *table-driven* machine and performs no calculation. The calculations, of course, have been carried out in the past and the answers preserved, but they could have been done in any way available, by abacus, by hand, or by mechanical adding machines.

It is evident that biological systems resemble the T machine, and not the P; they are *table-driven* computing devices. The values in the tables are stored in the sequence of the DNA, and were computed by evolution. Biological systems have ways of retrieving these values and using them. Thus, in our bacterial system example, the values stored in the genome for each enzyme might include: the amount of the enzyme produced, the affinity constant for its substrate and the parameters that determine the rate constant of the chemical transformation it catalyses. All of these are properties of the enzyme itself. Now I want to argue that if I had all of these values, I could set up a model of the entire system and it should work in exactly the same way as the real bacterial system. It should generate the same steady states of small molecules, and should enable us to make predictions in response to changes that can be reproduced in the real case. Since it would accurately represent the bacterium we could follow subsystems separately to gain a deeper understanding of the whole system. The theoretical model thus becomes an experimental platform for further research.

We notice that the global properties of the real life system are 'computed' by the network itself and we will be doing the same thing with the artificial network. We do not have to talk about these properties as emergent but look on them as arising directly from the properties of the components and their interactions. The whole is some special mathematical function of the parts, but we can safely say that this function is not the sum.

We can look at how this would fit into a complete reduction of cellular function to gene sequences. Again, we pursue our example of bacterial biosynthesis. We can certainly obtain the total DNA sequence of a bacterium — it has already been done for several microorganisms. From the DNA we can find coding sequences and translate them into amino acid sequences. We would now have to fold these up and then deduce from the structure of the folded protein what substrate it binds, with which affinity it binds it and what chemical transformations it catalyses. This would give us the numbers we need to go on to build our model. I believe that this task is impossible because we are trying to use genes to discover and analyse chemistry. The relationships are not unique, and two completely different enzymes can have the same substrate because they each can look at it in a different way. The computational requirements for simulating the folding process are very large and we may have to remain satisfied that we can do it in only a few cases. Indeed, if any of these questions could be shown not to be computable in the general case, we would have a formal rejection of reductionism, similar to that produced by Gödel for formal mathematical systems. However, it is certainly impractical to do it this way now, and in order to pursue the main question we would simply determine the values we require by measurement. Note that if we only had the DNA sequence we could still find the values by expressing the sequence in bacteria or we could chemically synthesize the proteins, let them fold in a test tube and measure the affinity constants and other values there. You could even consider the latter as an analogue device that uses a special kind of parallel computation to determine the numbers we wish to know.

We have sketched in outline how we might proceed from biological computation to computational biology. Because the genes can only specify components of the system, this linearizes complex systems. Furthermore, because the numbers are set by an evolutionary process, the system can grow in complexity by accretion; new components can be added, and new interactions can be tried. All of these are the products of simple mutational changes with natural selection taming the anarchic demons that do the work. The problem of emergent properties is avoided by showing that the parts when placed in the correct global environment will automatically compute the whole, and this would provide answers to questions of coherence, organization and integration which many biologists find puzzling today.

It is important to emphasize again that all of the systems we study are the products of evolution. That many of them resemble each other is the result of common origins and not of unique design. There may well be many different ways of getting to similar ends and we need to get away from the law-based explanations of physics. Furthermore, our task in explanation should be limited to the question of the effective construction of such systems and not to one of their design. Biological systems are not designed; that is the difference between natural engineering and man-made engineering. Unfortunately, the man designing a chemical factory cannot afford the luxury of spending a billion years evolving it; we probably have nothing to learn from him but he might well be able to learn from us.

Finally you will note that our computational models do not contain more than the real system. In fact, I believe it will be necessary to have special programming systems that prevent the programmer from imposing on the model his ideas of how it ought to work. With the advent of computing to biological research, there is often the thought that we could build computing systems that would themselves provide us the answers to the problems we are facing. It seems to me that this combination of artificial intelligence and human stupidity is the wrong one; what we need to combine is human intelligence with artificial stupidity.

Reference

Brenner S 1981 Genes and developments. In: Lloyd CW, Rees DA (eds) Cellular controls in differentiation. Academic Press, London, p 3–7

DISCUSSION

Williams: First of all, what is the K_m you use? You described it as a property of a particular protein even inside a cell, but it isn't—it becomes the property of a system inside a cell. The K_m is a constant if you measure a single enzyme reaction rate, but if you have the enzyme in a cell, which is coupled to another reaction system, and then this enzyme is coupled to others, the 'K_m' of the enzyme is linked to feedback and feed-forward concentrations of many substrates. So you have a K_m which should be written down as a variable of a string of properties. This will alter the order of the problem. A single enzyme can be treated by linear reaction kinetics, but in a controlled pathway its rate has higher-order terms and is non-linear.

The reason for this can be seen in the problem of an electrical device with feedback, except in the case of enzymes in cells the feedback uses mechanics. It has an allosteric communication system which adjusts K_m and V_{max}, and the number of chemical concentration terms which may be involved in the overall feedback may be as many as 20. The computational problem that the description of these particular protein kinetics may meet as a consequence of this is one of enormous complexity. When you try to work this out as a multiplicity of linear problems, I believe you will fail utterly.

Brenner: You tell that to *E. coli*!

Williams: I'm not like you: I can't speak directly to *E. coli*!

Hess: I do not agree with the interpretation of Bob Williams with respect to the global significance of the K_m or a more complex allosteric constant of an enzyme. Both constants are expressions of the reaction mechanisms and structure of individual simple or allosteric enzymes and nothing else. In a model of the dynamics of coupled enzymic reactions, such as glycolysis, the K_m and allosteric constants of all enzymes involved enter the flux equations which express the coupling of the fluxes of all enzymes in a reaction pathway in sequence and/or in feedback–feedforward loop pathways, whenever the process is activated by proper substrate input. Generally, because of the non-linearity of the rate laws of allosteric or interconvertible enzymes, complex dynamic regimes, such as oscillations, evolve. More than 30 years ago, together with David Garfinkel, I lumped the individual enzymic reactions of glycolysis together in more than 120 differential equations based on the principle given above and yielded a perfect fit of the steady-state behaviour of the process. This concept has been applied to many other processes in the meantime, showing that the principle of reductionism holds and allows us to resynthesize the dynamics of the process by mathematical models.

The question of how enzymic structures were selected in evolution, which display their function with proper turnover numbers, K_ms or allosteric constants to fit the cellular requirements for an efficient and optimal process, is another very interesting topic. No doubt, that in a process like glycolysis the fit is as such that the concentration of enzymes and of enzymic intermediates and controlling ligands under steady-state conditions fully meets the dynamic requirements of this most important energy-supplying system in living cells.

Garcia-Bellido: One of the reasons I liked your paper is because it offers an explanation for much higher levels of complexity. Development results from local effects, and there is no brain or mysterious entity governing the whole: there are local computations and they explain the specificity of something that is historically defined. The affinity constant is defined by experience accumulated through millions of years. The question is, what does it mean to explain things, to abstract them and to go to the essence of the phenomena? This is now a policy issue: what shall we do? Presumably the profound logic is something like that but without the details. And now we come to this classical question, is God in the details or is God in the logic? But if it is a way of handling problems then the next question is what should we look at?

Brenner: It seems that the approach taken by Dennis Bray is exactly right: that is, to take well-defined finite systems and explore them in detail, to facilitate detailed matching of prediction with experimental observations. It is important to know as much as you possibly can about the system, because sooner or later you will have to answer the 'other wire' sceptic who will ask you how you can be sure there is not another pathway responsible for the effect you are studying. If you can say that you know all the wires, you can start to build models and test predictions directly. That's the only way we will be able to handle the complexity of signal transduction, for instance.

Garcia-Bellido: In order to what? To cure cancer, for instance? The logical aspect of signalling is already understood in the profound sense. Another thing required by clinicians is to be specific in probing a particular problem with a particular receptor when someone has a colon cancer, for example.

Brenner: I want to know how these things came about to be the way they are. The fundamental problem we have is how these systems evolved. I want to know how they got to this level of sophistication.

Maynard Smith: I want to ask a fundamental question. *E. coli* only knows what its genes have told it, plus some relatively boring information about its environment. By and large we know that what the genes tell it is how to make proteins, and the laws of physics tell these proteins how to fold. What the proteins provide are the rate constants of the enzymes — what you call the numbers. In order to understand a biochemical problem, you need to know the K_ms and other constants of the enzymes. Is it true that all you have to know is the structure of a particular protein, or does the number depend upon all the other proteins in the cell? After all *E. coli* does know what all the other proteins are.

Brenner: In answering that question, I'd first like to describe an important experiment, which has preoccupied me for many years. If *E. coli* is placed in a medium made with heavy water there are a few hours when growth stops and then when the culture emerges from this state the bacteria grow in the steady-state but three times more slowly. There is three times more RNA and protein per DNA than is found in protonic medium. If this culture is maintained for 1000 generations we can ask whether there is any adaptation to the deuterium medium. I did this to try to evolve a bacterium that would be the final safe organism for genetic engineering: that is, it wouldn't grow anywhere except in my laboratory because D_2O did not exist anywhere in the universe until it was made by Urey in Chicago in 1936. How do we explain the first adaptation? This isn't something that *E. coli* can have any memory for, because previously there was no D_2O around. In a sense that's already an interesting experiment, because it shows the system can adapt to something completely unknown, that is, it has some generality. In brief, we can explain what is happening as follows. Most of the chemical reactions proceed at the same pace in heavy water except for a few. These are those in which hydrogens are removed, for example in the reduction of succinic acid to fumaric acid. Such reactions can be slowed 10- to 100-fold. The flow through a pathway containing such a step will be reduced and, automatically, feedback inhibition will be released. If that is not enough, there will be a de-repression of the synthesis of these enzymes to get more enzyme to make more stuff. Making more enzymes requires more machinery, and that is why there is an increase of RNA and protein relative to the gene.

However, everything is confounded by the fact that quite often a whole pathway is under one control. Thus, when more of the rate-limiting enzyme is required, everything is de-repressed. In the hydrogen system, the levels of the enzymes have been adjusted and normalized to the protonic universe. In the deuteronic universe this balance is no longer valid. Thus unnecessarily excessive amounts of some enzymes are being made. Since protein synthesis is expensive, we can ask whether the bacteria will improve in D_2O and, if so, how. This is a beautiful system, because every step will count, as long as it reduces the load on protein synthesis. It will be exciting to run this evolution experiment and then find out what happened by sequencing the genomes of the improved variants. This is not too far fetched: technology will soon be available that will be able to sequence a genome in a day.

Williams: That is a good experiment, but the main effect of D_2O is not on the proton reactions, but on the free radical reactions. DNA is made in a free radical step which has a 20-fold deuterium versus hydrogen dependence.

Kerszberg: Your proposal relies on the enzymes themselves as analogue computers to provide the affinities. You want to put the affinities in a digital computer: I would submit that you might as well leave them in the cell and use the cell as an analogue computer of itself: what will you have achieved?

Brenner: That is an interesting proposal, because someone will ask how we are going to integrate all of this. That is the wrong question: the cell integrates this. We should ask how we are going to understand how the cell integrates this. If you want to have it integrated, just use the organism itself.

Kerszberg: But isn't that more or less what you are doing?

Brenner: My organism is calculating the effect of D_2O without bothering with all the details, so that's true in that sense. But I want to understand how it does it.

Wolpert: I must confess to being a little confused. Are we saying that we will never be able to compute the cell?

Garcia-Bellido: You cannot compute the cell because it is a historical device. You would have to compute from the very first molecules interacting with each other. The problem is not that we cannot describe it.

Bray: Anybody who deals with multiple biochemical reactions knows that the outcome is very sensitive to the rate constants and the K_m, and unless you can specify them exactly, you won't know how it works. In a sense that's depressing because it means that you're going to have to know all these rates in the cell in order to understand how it is working. But there's another way of looking at it: a few years back we did something similar to the experiments proposed by Sydney Brenner, except that we used a computer (Bray & Lay 1994). We built a small signalling pathway which had an input (a ligand) and an output (a phosphorylated protein) and we gave it seven reactions. We tried to train that network to give a certain input/output relationship. So, for example, we gave it a step increase pulse of ligand and we asked the network to invert it to produce a negative pulse of the output model, or we asked it to give an optimal response in certain concentrations. And then what we did was to simulate evolution in a very crude sense by making random changes in the rate constants of all the seven reactions. We generated families of networks and selected those that were closer to the desired output. If we did this enough times, of course, we came up with a network that performed in the desired manner. We hadn't put the K_ms and rate constants in; instead, they were determined by the boundary conditions that we applied. It seems to me that this is what must happen in the cell. Evolution applies the boundary conditions and the rate constants just sort themselves out.

Nurse: Did you get more than one solution? Were there different ways of connecting the network to give you the same biological response?

Bray: Our circuitry was defined. In broad terms, the solution was always the same, every time we did it. It made one high affinity receptor and one low affinity receptor and they had distinct binding and catalytic rate constants. Perhaps if you have more than seven reactions there would be more scope for multiple solutions. I should add that although the solutions were always broadly similar, they were never identical. There was always variation in the precise values of rate constants arrived at by the computer — but this may be due to the practical consideration that the optimizations were of finite duration.

Holmes: A cautionary remark: before one starts trying to model eukaryotes, one should remember that eukaryotes invented vectorial chemistry. A good example of this is neuronal transport. Kim Nasmyth recently published two papers on the A-α mating types of yeast (Bobola et al 1996, Jansen et al 1996), which you would have thought is a simple system. His analysis showed that the determining factor is in fact a myosin, and it turns out that the whole thing is controlled by vesicle transport into the growing bud. I don't know how you are going to get that out of the simple-minded models.

References

Bobola N, Jansen RP, Shin TH, Nasmyth K 1996 Asymmetric accumulation of Ash1p in post-anaphase nuclei depends on a myosin and restricts yeast mating-type switching to mother cells. Cell 84:699–709

Bray D, Lay S 1994 Computer simulated evolution of a network of cell-signaling molecules. Biophys J 66:972–977

Jansen RP, Dowzer C, Michaelis M, Galova M, Nasmyth K 1996 Mother cell-specific HO expression in budding yeast depends on the unconventional myosin myo4p and other cytoplasmic proteins. Cell 84:687–697

Reductionism in learning and memory

W. G. Quinn

Department of Brain and Cognitive Sciences, Massachusetts Institute of Technology, Whitaker College, Building E25-436, Cambridge, MA 02139, USA

Abstract. This chapter examines the successes and (at least for now) failures of reductionist approaches in dealing with the problem of learning and memory. Beginning with the work of Pavlov on classical conditioning and the theoretical work of Hebb, the paper traces the contributions made by studies on *Aplysia*, *Drosophila* and long-term potentiation in the mammalian hippocampus.

1998 The limits of reductionism in biology. Wiley, Chichester (Novartis Foundation Symposium 213) p 117–132

In this paper I want to examine the successes and (at least for now) the failures of reductionism in dealing with the problem of learning and memory storage. I'm going to talk mostly about historical work other than my own, although, like all birds I can learn other songs but I learn my own species' song best.

Reductionists work with simple working models of how complicated phenomena can be derived from more basic phenomena, and they believe that nature is simple at heart. One can be a reductionist whether one works from bottom up (inferring outside principles from inside principles) or top down, as has to be done in the psychology of learning. In the top–down approach, there is a distinction between the study of simple models and properties, which I would define as reductionist, and the black box approach, which can be useful but is not reductionist. The other distinction that may be worth drawing is one between active, hypothesis-driven science and passive approaches such as high temperature superconductivity studies, genetic trawls for mutations and biochemical purification of important components. These approaches are less reductionist but they can be valuable.

There are triumphs in the history of learning, and they started with the work of Pavlov. In his classical conditioning paradigm he effectively ignored the complexity of the dog and its behaviour, and studied isolated parameters as if the dog had only two synapses. Pavlov showed that the salivation response of the dog to the sight of food is not a reflex but is a learned response. This is a completely reduced system: artificial reduction is a useful tool.

The next major advance was theoretical. Hebb was a psychologist and a theorist, but a simplifier rather than a turbidifier. His approach was to acknowledge that the dog is a complex ensemble of many neurons, but then to look for the simplest synaptic properties able to give associative learning. Notice that associative learning — a change in behaviour when an animal experiences two stimuli roughly simultaneously — implies some sort of synergism in the dog. When the dog sees the food, a circuit has to make a connection to pre-motor neurons that lead to the salivation reflex. When the bell rings, it goes to the ear and triggers another circuit. The dog learns faster than it can build new synapses, so there must be a latent connection, but before the training it doesn't work. After training the bell alone will give the salivation response, so the connectivity changes. Hebb asked what sort of properties could do this, and of the possibilities, he chose the simplest, which is that the synapse involved is strengthened if the presynaptic cell and the postsynaptic cell fire together. This work was carried out in the late 1940s: recently this phenomenon has been found in synapses in the hippocampus, and these properties are actually encoded by a molecule, which I will return to later in this paper. Thus Hebb's work has turned out to be a triumph, albeit a delayed one.

The other outside-in approach is the approach that treats memory in animals as analogous to memory in computers. In computers the problem of memory is divisible between the problem of reading in and out ordered arrays of information from the environment (an imaging and circuitry problem), and the local mechanism that a particular piece of memory uses to store a bit of information. Therefore dissociation of global and local problems is something that Hebb was thinking about and which we have been thinking about ever since. In the brain there has been little progress made on the imaging and circuitry problem, but a fair amount of progress has been made on the characteristics of the local chemical tricks by which neurons change synapses and other properties.

In the study of learning and memory, there are four big systems used with varying degrees of reductionism. One of these is *Aplysia*, which has successfully been studied by Kandel and others (e.g. Bartsch et al 1995, Bailey et al 1996, Hedge et al 1997). A simple behaviour, the gill withdrawal reflex, has been studied extensively. Because this animal is simple, it is possible to trace the neuronal circuits and correlate the behaviour of individual synapses strengthening and weakening with the withdrawal reflex. This is a reduced system which is further reduced by taking the relevant cells and putting them in culture. If you study the system carefully you find out that there are a number of 'semi-swindles' and reductions that are only partially appropriate. The take-home lesson is that the short-term memory is encoded by changes in a second messenger system, cAMP, and longer-term changes are encoded by cAMP-dependent transcriptional activation. None of those statements are completely true: nevertheless, this is the model that has driven the field and the oversimplification has been justified in every case because of the unbelievable predictive richness and explanatory power it gives. Limited swindles are part of all the work in studying learning.

Learning in *Drosophila* was first studied in Seymour Benzer's lab by myself and others (Quinn et al 1974, Dudai et al 1976), and has continued since in my lab and others (e.g. Yin et al 1994, Feany et al 1995). *Drospohila* has 100 000–300 000 neurons: it is not a simple system, but at the time it seemed to be genetically the most convenient system for this work. Strong learning in wild-type flies can be demonstrated by an olfactory discrimination task in which the fly is exposed to two odours in air currents. Flies receive electric shock pulses on the first odour but not the second. To test for conditioned odour responses, the flies are then transported to a T maze choice point between converging currents of the two odours, where 95% will avoid the shock-associated odour (Tully & Quinn 1985). Thus *Drosophila* demonstrates real associative learning. We used this task plus genetic mutagenesis to isolate a number of informative mutants. *dunce*, isolated in Benzer's lab by Byers, is a cAMP phosphodiesterase (Dudai et al 1976), *rutabaga* is an adenylate cyclase (Duerr & Quinn 1982) and *amnesiac* codes for a neuropeptide that is homologous to mammalian pituitary adenylate cyclase-activating pathway (Quinn et al 1979). By using transgenics we showed that cAMP-dependent kinase and the kinase-activated transcription factor CREB are involved, which wasn't a complete surprise. One surprising result, however, came from an experiment done by Tully and Yin: if you flood flies with activating CREB by putting it on an inducible promoter, these flies have a flash-bulb memory (Yin et al 1995). They remember as well after a single training trial as normal flies do after repeated trials at intervals. Therefore the good news from flies is that you can go directly from genes to complex behaviour via mutations. The bad news is that you go from genes to complex behaviour without really understanding the subtleties of the intervening physiology and organizational network. It is shameful that we treat the flies merely as biochemical soup. Nevertheless, as far as figuring out the cellular pathways by which neurons encode permanent or semi-permanent changes, this information is a lot better than no information at all.

As we move up the pathway the simple systems approach works on the idea that what is true for *E. coli* is true for mammals. This is not completely correct: what is true for mammals is not necessarily true for *E. coli*. So there is a limit to what one can learn by addiction to simplicity. It turns out that when you study synaptic modification in higher organisms it looks different: it is NMDA receptor-based. However, if you look at long-term synaptic changes in the hippocampus, they seem to require cAMP-dependent protein kinase and probably CREB. It is likely that you won't learn everything from the simplest system you can get your hands on. On the other hand, in evolution, nothing is thrown away. Mechanisms aren't evolved in elegance and then tossed out: they are embroidered upon. Thus simple systems won't give the whole story but will give you relevant information.

The other story that is worth talking about is long-term potentiation (LTP) in the rat and mouse hippocampus, which was first studied by Tim Bliss in the UK (e.g. Williams et al 1989, Thomas et al 1994) with notable contributions from Graham Collingridge (e.g. Bortolotto et al 1994, Clark & Collingridge 1995). Lesioning the hippocampus produces memory deficits, and in region CA1 there are synapses that are strengthened by simultaneous presynaptic and postsynaptic activity. This turns out to be due to a particular molecular receptor: the NMDA receptor, which is ligand and voltage gated and responds only when there is transmitter from the presynaptic side and depolarization at the post-synaptic side. It conducts in Ca^{2+}, everyone's least favourite second messenger (because it sticks to stuff and is difficult to measure). In this particular case there is a nice run from behaviour down to a gene product and back to behaviour. There is a tissue-specific NMDA knockout from the Tonegawa lab that blocks LTP in this region (CA1) and the mice have impaired spatial memory (Tsien et al 1996). However, although we know that the hippocampus is important for learning, no one knows exactly how. The hippocampus is necessary for encoding declarative memories, but after a while these are stored elsewhere in the cortex. A lot is known about hippocampal circuitry, but how it helps the cortex store memories is a mystery.

Experiments in monkeys and lesions in humans have given insight to the other analogy from computers: memory storage. The prefrontal cortex stores working memories which are similar in nature to the RAM of the computer in the sense that they are dynamic, trial-specific memories. Other parts of the brain store longer-term memories. One is left with knowing things about the synaptic changes in the hippocampus yet being clueless about the other problem of exactly how complex memories are encoded. The second problem of computers — reading in and out ordered information from the environment — is not understood at all at the local neuronal level. This doesn't mean that the reductionist approach won't work, but reductionism has treated the problem like the USA treated Vietnam: declared victory and withdrawn. We'll have to wait for new techniques to make further progress.

There is one other application of the reductionist approach that is completely irrelevant but is sort of interesting, and that is to writing poetry. Shakespeare only wrote down once in one of his plays what he was doing when he was writing it, and this was in *A Midsummer Night's Dream*. His description of what poets do is completely reductionist, and completely selective:

> The lunatic, the lover and the poet
> Are of imagination all compact:
> One sees more devils than vast hell can hold,

That is, the madman: the lover, all as frantic,
Sees Helen's beauty in a brow of Egypt:
The poet's eye, in fine frenzy rolling,
Doth glance from heaven to earth, from earth to heaven;
And as imagination bodies forth
The forms of things unknown, the poet's pen
Turns them to shapes and gives to airy nothing
A local habitation and a name.

References

Bailey CH, Bartsch D, Kandel ER 1996 Toward a molecular definition of long-term memory storage. Proc Natl Acad Sci USA 93:13445–13452

Bartsch D, Ghirardi M, Skehel PA et al 1995 Aplysia CREB2 represses long-term facilitation: relief of repression converts transient facilitation into long-term functional and structural change. Cell 1995 83:979–992

Bortolotto ZA, Bashir ZI, Davies CH, Collingridge GL 1994 A molecular switch activated by metabotropic glutamate receptors regulates induction of long-term potentiation. Nature 368:740–743

Clark KA, Collingridge GL 1995 Synaptic potentiation of dual-component excitatory postsynaptic currents in the rat hippocampus. J Physiol (Lond) 482:39–52

Dudai Y, Jan YN, Byers D, Quinn WG, Benzer S 1976 *dunce*, a mutant of *Drosophila* deficient in learning. Proc Natl Acad Sci USA 73:1684–1688

Duerr JS, Quinn WG 1982 Three *Drosophila* mutations that block associative learning also affect habituation and sensitization. Proc Natl Acad Sci USA 79:3646–3650

Feany MB, Quinn WG 1995 A neuropeptide gene defined by the *Drosophila* memory mutant *amnesiac*. Science 268:869–873

Hegde AN, Inokuchi K, Pei W et al 1997 Ubiquitin C-terminal hydrolase is an immediate-early gene essential for long-term facilitation in *Aplysia*. Cell 1997 89:115–126

Quinn WG, Harris WA, Benzer S 1974 Conditioned behavior in *Drosophila melanogaster*. Proc Natl Acad Sci USA 71:708–712

Quinn WG, Sziber PP, Booker R 1979 The *Drosophila* memory mutant *amnesiac*. Nature 277:212–214

Thomas KL, Laroche S, Errington ML, Bliss TV, Hunt SP 1994 Spatial and temporal changes in signal transduction pathways during LTP. Neuron 13:737–745

Tsien JZ, Huerta PT, Tonegawa S 1996 The essential role of hippocampal CA1 NMDA receptor-dependent synaptic plasticity in spatial memory. Cell 87:1327–1338

Tully T, Quinn WG 1985 Classical conditioning and retention in normal and mutant *Drosophila melanogaster*. J Comp Physiol [A] 157:263–277

Williams JH, Errington ML, Lynch MA, Bliss TV 1989 Arachidonic acid induces a long-term activity-dependent enhancement of synaptic transmission in the hippocampus. Nature 341:739–742

Yin JC, Wallach JS, Del Vecchio M et al 1994 Induction of a dominant negative CREB transgene specifically blocks long-term memory in Drosophila. Cell 79:49–58

Yin JC, Del Vecchio M, Zhou H, Tully T 1995 CREB as a memory modulator: induced expression of a dCREB2 activator isoform enhances long-term memory in Drosophila. Cell 81:107–115

DISCUSSION

Gray: With regard to our main theme of reductionism, there are at least three different issues being confused here. Two of these were in this paper: first, the notion that you're being reductionist when you bring things into the laboratory (as Pavlov obviously did with his studies of dogs) and, second, the notion that one is being reductionist when one goes down to a simpler organism (e.g. from a human, to a dog, to *Drosophila*). I don't think that either of those two notions in any way poses problems of conceptual understanding of what the limits of reductionism might be — they're both aspects of basic scientific method. The one that's of interest here is whether you can reduce from one set of phenomena to another set at a more basic level. There, I think the study of learning and memory is interesting, because it shows that both the reductionist and the non-reductionist approaches are extremely valuable.

It's obviously valuable to do the kind of work that we've heard about in which you look in *Drosophila* for genes that are involved in memory and in *Aplysia* for changes in neurotransmitter function involved in forming an associative bond. Not only is it useful to study these mechanisms at those levels, but also some of the results do indeed generalize up to mammals. You can get a lot of extremely valuable information through the reductionist route in the sense that you're looking for those bits of the nervous system which change in order that learning can take place and memory can be stored.

But it is equally rewarding to study learning at the level of concepts about learning, without trying to reduce it further. One is able to say with some certainty, for example, that the principles of associative learning that Pavlov discovered, simply by pairing for example food with bells, apply to *Aplysia* and humans, because classical conditioning has been shown to exist in both these species, although it may not happen according to the same underlying neuronal machinery. The laws that Pavlov described for associative learning have proved to have enormous generality across many different species, and therefore they must reflect something that is built in not only to the nervous system in some specific way, but also to the transactions that organisms have with their environment, so that in all those cases associative learning has been a valuable thing for evolution to have led to, not necessarily always by the same route.

So, to go back to the main theme of our meeting, the study of learning shows extremely well that there is no conflict whatsoever between reductionist and non-reductionist approaches.

Quinn: It's possible that in making these distinctions I'm confused about reductionism, but it strikes me that some people have broad definitions, others have narrow definitions, and it would be worth making those outlines explicit so we can at least talk about categories. There is complete reductionism and there is limited reductionism: Pavlov was reducing as far as he could. Second, just because two inter-trial interval relationships are encoded in different parts of the brain doesn't mean that they're encoded via different molecular mechanisms. There's always the possibility that the broadness of Pavlov's transactional inter-trial interval temporal relations underlies the fact that they are being encoded by a countable number of molecular processes.

Rose: There is a degree of biochemical parsimony. It may not be embedded in the NMDA receptor, but I suggest it is embedded in the machinery within the cell. But that's not the crucial point: the crucial point is partly to distinguish between methodological reductionism — that is, how you do an experiment in the laboratory — and whether if you actually reduce a preparation in that particular sort of way or you understand molecular mechanisms of synaptic plasticity, you then understand the phenomenon of memory. We're all engaged in methodological reductionism because it's the only way we know how to do experiments, whereas dealing with the complex systems that most of us have to handle if we are interested in behaviour really does present enormous problems. But we do have problems that you partly identify in the context of the whole field of learning and memory. There is clear-cut evidence, for example, that comes out of the approach that you didn't describe — the imaging approach — but is hinted at in what you said about the hippocampus, that learning takes place in one region of the brain and is then manifested during recall in other regions of the brain. We are therefore dealing with dynamic systems, not fixed systems. The fact that we can identify synaptic changes occurring when learning and memory formation take place doesn't mean that memory is therefore embedded in those synaptic changes, because memory is a property of the organism, and is only manifest at the level of the system of behaviour of the organism. Consequently, if you want to understand memory, you have to understand both the reductionist approach of what's happening mechanistically at the synapses, and also what that phenomenon means at the level of the organism. Learning and memory is only a meaningful concept at the level of the organism and its interactions with the environment. It doesn't make any sense at the level of the synapses that you and I study.

Quinn: To my mind, that makes it all too vague and inaccessible.

Barlow: For some purposes it may be better to describe what happens when an animal learns in terms of the probabilities and conditional probabilities of events in the outside world and the animal's behaviour, rather than in the reductionist language of molecules and modifiable synapses. Such a probabilistic account can provide a shorthand from which one can make useful predictions, without getting lost in half-understood possibly irrelevant detail. It would be reductionist in the sense that it simplifies and 'economizes thought' (see the last section of Barlow

1998, this volume), but not in the sense that it reduces to physical or molecular mechanisms.

Quinn: That doesn't seem non-reductionist; it just seems evolutionarily driven, and then one selects appropriate mechanisms.

Perutz: A simple biochemical question: are the three enzymes involved in short-term memory part of one cascade?

Quinn: Yes.

Perutz: Are you knocking out consecutive steps?

Quinn: Yes. There's this first-order kinetic sense that comes directly from the molecular biology, which is that if you want to encode neural changes relatively rapidly but more persistently than the duration of an action potential, second messenger systems seem like a convenient way to do it. The brain is full of them. It turns out that if you want to encode them slightly longer than that, neuropeptide activation of second messengers has more duration. So the fact that a middle-term memory mutant fits in that way makes good sense, although you don't know that it's going to be that pretty when the smoke clears. If you want to encode more enduring changes, second messenger-activated gene transcriptional events are a good way to do it, which is what led to the CREB idea. So there's this simple pretty sequence of kinetic events which is going to get more complicated and partly fall apart, but it's better than nothing.

Rose: It is more complicated than that in a crucial way. That is, you can bypass or counteract the effect of the knockout by changing the training procedure in many of these cases, and you can then reconstitute the learning.

Raff: One of the successes of reductionism in studying learning and memory has been the demonstration that there are structural changes that occur in synapses in long-term learning. This is a major advance.

At some point we should discuss what molecular reductionism might tell us about learning and memory in ways that we couldn't possibly have predicted. New neuroanatomical components that one had no idea about are being described simply by looking at where specific proteins are distributed in the brain. My guess is that the reductionist approach, even where it is just a fishing expedition, will lead to real understanding in unpredictable ways, and that the molecular and cellular basis of memory, learning and other higher brain function could well emerge bit by bit, until the mystery gradually disappears, just as has been happening in developmental biology.

Wolpert: I'm curious that you didn't mention 'neural networks' in your paper.

Quinn: There are a number of reasons why I didn't talk about neural networks. This approach is going to be necessary in understanding complex behaviour, but it strikes me at this point to be premature: the people that do it are honest, but it hasn't yet influenced our actual understanding of visual processing, for instance. Another thing about neural networks that bothers me is that people say neural networks can learn, but it always struck me that what they were doing was optimizing some goodness of perception, and what they call learning is the fact that they don't work very fast, and so it takes many cycles to optimize, so they learn over time. I think this is a necessary approach, but so far it hasn't contributed much to our understanding of the real problem of neural imaging in biological systems.

Bateson: Some of the neural networks have been used not for biological explanations, but because people wanted smart machines. These have been developed by people who are interested in artificial intelligence. Such networks are not helpful in biological explanation. However, I think that neural networks that are biologically plausible at both the behaviour and neural levels do help us to understand the systems properties of behaviour.

Morgan: I agree that neural networks have told us nothing about vision. I also agree with what Horace Barlow hinted: that in many cases all they amount to is optimized statistical classifiers. A whole class of them is logically equivalent to principal component analysis, so it's not surprising that they can learn to classify the world into tanks versus non-tanks, for example, which is the sort of thing they are good at doing.

I should try to say what a cognitive psychologist might say about neural networks, because they find them extremely useful and think that they have illuminated topics such as memory. Models of reading, such as that of McClelland & Rumelhart, have many adherents, and they certainly can do striking simulations, particularly of the effects of brain damage. If you teach the network to read and then make arbitrary lesions in it you get a bewildering variety of defects, but amongst them are some that look like deep dyslexia, which is the phenomenon that you get clinically with people who read and make complete semantic confusions, so they might read 'sheep' as 'lamb' for example, showing us that the information is being processed up to a semantic level quite clearly, but something is missing.

Mitchison: I wanted to mention Christoph von der Malsburg's work (von der Malsburg 1973). He showed that you could construct a kind of neural network which doesn't have a specified target output but just has a Hebbian learning rule. He simulates nicely the appearance of ocular dominance columns in the visual cortex. This has been helpful in showing that the Hebbian rule is sufficient to explain something that otherwise seems rather puzzling.

Barlow: The trouble with neural networks is they were originally based on the Pitts/McCulloch neuron, which was too simple. The technique they developed has been applied for example by Dennis Bray to help understand enormously complicated internal networks. But what that tells us is that when the connectionists were developing neural networks, they thought you needed a dozen neurons to explain things we now know can be done by the interactions between half a dozen internal signalling molecules. They invented a new style of computation, but they did not succeed in making convincing models of what they originally planned to model, namely the production of psychological behaviour, because the elements they used were not good abstractions of real neurons.

Quinn: I said I thought neural modelling was necessary to the approach, but it's going to look premature and irrelevant until it has a triumph, and that hasn't happened yet.

Ashmore: There's clearly a linking issue here, which is the actual structure of the system. We have talked about neural networks and the molecular basis of memory, but the question is: how do these map onto the real organism?

Quinn: I think that Martin Raff has given the answer: if you know that surface molecules are induced for long-term changes and you have antibodies to those surface molecules, you can figure out all kinds of ways that you could stain activated sets of neurons.

Raff: But the reason that modelling may be premature is that molecular neurobiologists are continually discovering new subcompartments of the nervous system that we had no idea about. A region that was believed to be homogeneous turns out to consist of a hundred different subregions. If you don't know that and you're modelling the region as if it's homogeneous, is that really useful?

Kerszberg: Small nervous systems, such as the lobster somatogastric ganglion, have been simulated with great success. So has lamprey locomotion, to a great level of detail and with quite a bit of predictive power.

Gray: You can use neural nets to model the real nervous system for which you do indeed need to know the wiring, but you can also use these in the way that Michael Morgan was just describing, where you model a system which will correctly produce outputs given inputs. Here, there have been quite considerable successes, certainly within the general field of learning theory at the level of formation of associations and so on. For example, Schmajuk (1997) has just published a book in which he has used simple neural network modelling to account for a large number of phenomena in general associative learning, without directly referring to how the nervous system is actually made. Obviously, at some stage this modelling has got to make contact with the way the nervous system works, but it is extremely useful even now to show that with a relatively small set of assumptions you can account for a wide range of phenomena.

Raff: Even if, in reality, it turns out that it's done in a completely different way?

Gray: No, I agree with you, that in the long run, reality must be the acid test.

Raff: But how would you feel about a model that was satisfactory in explaining something, but it turned out that it isn't the way it actually happens?

Gray: I would suppose it impossible, if it was satisfactory in explaining something, that it wouldn't be able in some way to encapsulate certain principles which are relevant to the way the real nervous system works.

Dover: I have a question for Steven Rose concerning the whole organism approach. If you take the Pavlovian response, one could describe that as some sort of behavioural plasticity which you seem to say is not describable totally in terms of the events of synaptic events. Are you suggesting that there is something else out there which can explain that response?

Rose: The issue is that one can understand how the phenomenon of learning and memory *per se* is evolutionary important in the survival of the organism, and it is a property therefore of the organism as a whole. Why animals or organisms learn and remember is a question which you have to understand in terms of the history and evolution of the whole organism. Furthermore, it's abundantly clear that even to understand the mechanics of memory, you have to understand the wiring of the system, and not the individual synapses that are in it. So we can ask 'how?' questions, which are tremendously important, about the involvement of these cell-surface molecules and individual synapses. The moment you do that and try to study memory, you end up with paradoxes. For example, those regions of the brain which show the short-term changes associated with learning are not the same regions where the long-term changes occur in synaptic stabilization. Again, you have to understand the system, in order to understand the meaning of those synaptic changes that we can study so exquisitely in the laboratory. This isn't antireductionism, this is simply saying that we're dealing with different classes of questions and approaches. It is an epistemological pluralism I am advocating: we need them all.

Gray: Steven has already mentioned the fact that learning can be 'silent' until you come in later to test the system by a different environmental challenge. Let me spend a couple of minutes describing one specific instance of that general phenomenon. This again goes back to Pavlov, who described something known as sensory preconditioning. We do it regularly in our own lab. You can associate two stimuli, neither of which has any particular biological significance to the animal you are studying. For example, we start by associating a simple light followed by a simple tone. Subsequently we associate the second of those stimuli, the tone, with something that is biologically important, such as a footshock. The light has never been associated with footshock, but the animal becomes afraid of the light, because this is associated with the tone which is now associated with the footshock. As with classical conditioning, the sensory preconditioning phenomenon is ubiquitous. Supposing you did the first stage of that sensory preconditioning experiment: you associate light with a tone. Nothing that you could look at in the synaptic changes, which would undoubtedly have taken place as a result of that conditioning, would tell you how the animal will subsequently respond once the tone is paired with the footshock. You can, however, tell from the general knowledge that one has of associative learning, that the light will then produce a response as though the animal was afraid of the footshock.

These are not antagonistic approaches. There's a terrible tendency to think of reductionism and this other approach ('antireductionism' is probably the wrong way to describe it) as being opposed to each other. They are not; they're complementary.

Brenner: I wanted to comment on simulation. I believe you can separate the question of whether simulation is a description or an explanation by examining the programme. I would say it's an explanation if it is written in the 'machine language' of the object being simulated.

Morgan: That was exactly the point I wanted to take up following from Martin Raff's question about whether simulation of a learning process would be any good if it turned out that it was in real life this was done in a different way. We have to clarify here what we mean by 'in a different way'. This is actually rather a deep issue. If we stick to learning, to make this a little more concrete, suppose that a simulation were based on a formal computational model of learning. That instantiation will involve an algorithm, and that algorithm may be very different from the one that the brain uses. But is it doing it in a fundamentally different way? It depends what you mean by 'fundamentally different': if it incorporated the same basic computational theory, it might not be. Of course, what I'm just doing here is just echoing the distinction that David Marr taught us. He said that if you are looking at information processing devices, the distinction is between the computational level theory of what they're doing and the algorithms (which we know are totally different between different machines). Thousands of machines can do arithmetic but the algorithms that they use might be quite different.

Another simple example from vision is the stereo matching of images between two eyes. This is a fundamental problem. One can come forward with various algorithms for doing it which are successful. None of them may be the algorithm that the human visual cortex uses, but it is possible that they're all related on a deeper and fundamental mathematical level, in that they're solving a computational problem by imposing constraints. Those constraints might actually be the same in all the algorithms.

Therefore my question to both of you is: do you think there is any utility to this computational level of analysis? Would you say it's the algorithms that really matter if you're a biologist?

Quinn: I find the work of Marr on the cerebellum and the hippocampus highly relevant. In his model, the cerebellum is set up as a differential amplifier to measure the difference between the motion that you want to make and the motion that you actually make, so you do motor learning that way. There are a class of synapses in there which have to be modified in a certain way. He got the actual way wrong, but with Ito's data you can turn his model into a functional explanation for how the cerebellum might work. Furthermore, it is testable. Even if it's still wrong it gives you some framework in which to think about how the cerebellum is organized in reality.

Morgan: Is an explanation only an explanation when you have the machine code? Is a computational level of analysis just not biology?

Brenner: I'm talking about a relevant one, because otherwise it belongs to some other realm such as mathematics. I would like to see the simulation written in the machine language.

Raff: Let me give an example that might help answer your question. In the early days of immunology, when one thought there was only one kind of lymphocyte, there were mathematical modellers trying to explain immune responses, immunological tolerance, self–non-self discrimination, and so on. Then it was discovered that there were two classes of lymphocytes, each doing completely different things. Then there were three, then four, and so on. The original models were largely meaningless, because they were missing most of the components. In a simple reflex response where we know there's only a small number of types of neuron involved, modelling makes a lot of sense: in fact, you couldn't understand it any other way. But when there are hundreds of different types of neurons involved in a response, modelling them as if they're homogeneous is probably a waste of time.

Morgan: But the aim of the exercise would not be to model the neurons, it would be to produce an information processing level account of the tasks which are being faced and the problems which must be overcome.

Raff: Suppose you didn't know anything about synaptic inhibition and tried to model a response with just synaptic excitation. Would the models be useful?

Bateson: Some of the simulations with neural nets look very plausible, and even though they're modelling thousands of neurons, they can do things that are extremely life-like. Providing we use rules that are known to be implemented in the nervous system, there is a possibility that they will advance understanding of the normal basis of behaviour. If they do not, fine, then we have to move to something else.

Williams: Is the distinction that we are making between properties of systems which can be described by the identification of individual molecules, and those that only belong to systems of molecules? If so, then in the second case a system is not reducible. For example, if you were to take a molecule, you could describe its bonds, but you can't describe its entropy unless you deal with a system of many molecules. Entropy is not reducible to a property of a molecule. Thus *random* kinetic energy of a large number of molecules is described by temperature, which is very different from kinetic energy of a single molecule, which is a vector. There is a basic distinction between these collective properties and properties of single or pairs of units. It works out in the following argument. Let's take a property such as superconductivity. This exists only in a system, but there are various systems which show superconductivity of metals which do not work when they are used in an attempt to deal with the superconductivity of copper/lanthanide oxides. So superconductivity is a property of a system which arises and can arise in different ways in different systems, rather than belonging to a property of individual atoms or basic molecules. You cannot reduce the description beyond a certain level.

Wolpert: But you are nevertheless explaining superconductivity in terms of the behaviour of the electrons.

Williams: Not really; you can only explain superconductivity in terms of electrons in a collective mode and level if you explain the properties of electrons in the surrounds in which you find them — in a lattice. It is the lattice which decides the possibility of the electrons behaving in a certain way while the electrons carry the current.

Gray: This is analogous to the example I gave earlier of sensory preconditioning, in which the associative learning can only be fully described if you take account of the organism and its interaction with the environment. For any individual system, however, there must be some underlying synaptic changes which you should be able to isolate.

Kerszberg: Returning to the question of the usefulness of neural networks, Zipser & Andersen (1988) took the relationship of the inputs coming from the lateral geniculate nucleus (LGN) to the visual cortex and made a standard back propagation model. In this there is an input layer, an output layer and in between there are hidden units. The input–output relationship is imposed on the system by learning, but you don't say anything about what the hidden units might be doing. This sort of model is not biological at all — this algorithm is a purely mathematical contraption. Nevertheless, they correctly predicted the response of cells in the visual system that were supposed to detect combinations of certain features. Sure enough, such cells were later found to exist in the visual system, and other cells which never showed up in the simulations did not show up experimentally, either. I would suggest that sometimes the brain solves a mathematical problem with its own neural network-type methods, and although mathematicians may solve the problem in a completely different way algorithmically, because of the mathematical properties of the problem itself, ultimately the same sort of features might show up in both solutions. These features would be characteristic of the *problem* and not so much the biology.

Barlow: The hypercomplex cells were described long before that work, so it wasn't a prediction that was later fulfilled.

Nurse: We've touched on an interesting issue here, which is the use of metaphor and analogy in scientific understanding. Is it ever useful to think about things metaphorically? In other academic disciplines, metaphors are often used, and I suspect that scientists could make more use of metaphors in their thinking. To a large extent I agree with Sydney Brenner's notion that one can only understand a system if the same machine logic is used, but sometimes there are fundamental similarities in the way things operate, and metaphors may be useful in these circumstances.

Quinn: The metaphors that physicists use, for example, are simple and intuitive, and neural networks don't have that simplifying property.

Brenner: What David Marr did was different from everything else that went before, because he didn't start with any abstract theory of mathematical nets, but with the functional anatomy of the cerebellum. He asked: given this machinery, what can it do? He modelled in terms of the machinery, so he had provided something that could in his view be generated by what was there. Whether his theory was right or wrong does not matter: the point is that he could test it. This is different from having an arbitrary net that might or might not work in the same way as real life.

Maynard Smith: When I read Marr on vision, what I got from it was something very different from Sydney. If I understood him, what Marr was saying about vision was that any object that is going to see things — whether it is an organism or a machine — has certain problem it must solve, and there is probably only a finite number of different logical ways in which those problems can be solved. If you are playing squash, for example, you want to hit the ball and if like me you only have one eye there are real problems, and there are only certain computational solutions to those problems. This is hard to prove: in effect it amounts to saying I can only think of two ways of solving this, although there may be others.

Are behavioural scientists and neurobiologists ruling out as a useful enterprise the procedure of saying, look there really is only a small number of possible ways that this very difficult problem could be solved, so the brain has to be doing it in one of these ways, and let's have a look?

Brenner: Let me give you one concrete example, which is a fielder catching a cricket ball. There are two possible ways of doing this. In the first, he could make the observation as the ball left the bat, then compute the trajectory of the ball and calculate how fast and in which direction to run to intercept it. The other way is for him to ask himself a set of questions: for instance, if it is going to the left he will start running to the left. Therefore he would solve the problem by division and interpolation. I think it is done the second way and not the first.

Maynard Smith: Is it useful to ask which way the brain is doing it?

Brenner: Absolutely, and then what becomes plausible, is the plausibility of the machinery.

Gray: The remarkable thing about that or any similar example is that we know from experimental work that all of those computations, however they are done by the brain, are completed in a period too short for the conscious percept of the ball having ever left the opponents bat (Gray 1995).

References

Barlow H 1998 The nested networks of brains and minds. In: The limits of reductionism in biology. Wiley, Chichester (Novartis Found Symp 213) p 142–159
Gray JA 1995 The contents of consciousness: a neurophysiological conjecture. Behav Brain Sci 18:659–680
Schmajuk NA 1997 Animal learning and cognition: a neural network approach. Cambridge University Press, Cambridge
von der Malsburg C 1973 Self-organization of orientation-sensitive cells in the striate cortex. Kybernetik 14:85–100
Zipser D, Andersen RA 1988 A back-propagation programmed network that simulates response properties of a subset of posterior parietal neurons. Nature 331:679–684

Hearing

Jonathan Ashmore

Department of Physiology, University College London, Gower Street, London WC1E 6BT, UK

Abstract. The aim of this chapter is to describe some of the features of the processing of the auditory world and how different levels of explanation are appropriate to the understanding of hearing. The working of the inner ear is best seen as a the operation of a purposefully structured machine for the extraction of biologically meaningful components from a sound. Physical scales determine in large part the appropriate description of the auditory system.

1998 The limits of reductionism in biology. Wiley, Chichester (Novartis Foundation Symposium 213) p 133–141

Since the work of von Helmholtz in the mid 19th century, the study of hearing has been a field where physics, engineering and biology meet. Hearing research is pervaded by images and processes which are derived from engineered devices. There is structure to be seen at all levels in the organization of the inner ear and the most prominent view of hearing sees the cochlea as a machine. From there on, the subsequent processing of a complex sound by the nervous system and its determining role in behaviour is seen as depending upon the computational power of the nervous system. The limits of analytical approaches to understanding how the brain operates are set by the physiological techniques and the types of explanation that we wish to produce.

The analysis of the processes in the central nervous system has always been problematic for physiology because the techniques for investigating high level processing do not give much information about the component parts. Conversely, the emphasis of neurophysiology on the properties of individual neurons and local pathways is difficult to synthesize to a nervous system where many millions of cells are devoted to processing information. For this reason, the recent developments in functional imaging (in particular, functional magnetic resonance imaging, fMRI) have attracted considerable attention as a way of investigating, non-invasively, more complex cognitive tasks.

The purpose here is to discuss to what extent an understanding of the components out of which the cochlea is made (a) helps us to predict how the cochlea responds to biological sounds and (b) helps to guide us towards understanding its construction. It is often suggested that the study of specialized organs such as the ear teaches us few general principles since the cochlea consists of highly specialized structures which exist at the fringes of 'mainstream' neuroscience. However, many of the issues which face contemporary neuroscience also face hearing: the issues of how cells are specified, how they are assembled together, how molecular descriptions lead to cellular descriptions and how complex response patterns emerge all apply with as much force in hearing as in any other area of neuroscience. I hope to be able to convince you that describing the interaction between the parts of the auditory system is a non-trivial exercise but one which is aided by a particularly transparent set of structural relations.

What are the problems in hearing research?

Whether we hear normally or not is a health issue. This means that there are medical descriptions when explaining how hearing works that may often be no more than a list of prescriptions for treatment. Auditory psychophysics and its clinical cousin audiology, however, are concerned with (usually non-invasive) measures of hearing ability. Auditory physiologists, who work at the next level of hearing in both the peripheral and central nervous system, focus on a cellular description of hearing. Here, as in other areas of neuroscience, the questions become ones of how far the description of psychophysical performance (and in particular the sensitivity and selectivity of the auditory system) can be predicted from a description of the cells and their interactions. Finally, the growth of molecular methods of analysis has begun to impinge on the explanation of hearing processes, so the question we are beginning to be faced with now concerns how far molecules and assemblies of molecules within the cells of the cochlea can explain the cellular physiology.

A short tutorial on the inner ear

The first step in understanding hearing depends upon taking apart the cochlear machine. The cochlea is a fluid-filled compartment within the temporal bone, a situation which has made its study lag behind other sensory organs such as the eye. Its central dividing partition is the basilar membrane: sounds are funnelled into the ear canal by the outer ear, transmitted through the middle ear to the inner ear, and cause this partition to vibrate. The technical problems with studying hearing arise because the amplitudes of basilar membrane disturbances are in the order of nanometres.

The hair cells which detect and transform the incoming sounds are modified epithelial cells which are specialized to detect small deflections of the processes which project from their apical surface. The processes (stereocilia) are modified villi which rotate about their insertion points on the apical surface of the cells. In the process they undergo a relative shear movement (Hudspeth 1989). Because of the very small movements within the organ of Corti at threshold of hearing, the relative slippage at the stereocilial tips is measured in nanometres. Nevertheless, this movement is thought to be sufficient to gate the transduction channels located at the stereocilial tips and to allow ions to enter the hair cells. The evidence for this mechanism has drawn on a large number of different types of experiments, both structural and functional (recently reviewed in Dallos 1996). The direct gating of channels cannot be investigated directly except by means of electrophysiology. The structural basis for the gating is elaborate: it involves the coupling of the channel to the stereocilial motion by protein links and a tensioning apparatus to keep the link tight using a molecular motor at the end of the link.

A second mechanism which reveals the interface between mechanics and biology is sound amplification in the mammalian cochlea (Ashmore & Kolston 1995). This mechanism exploits properties of a specialized subclass of hair cell, the outer hair cells in the organ of Corti. They disappear progressively with age and this disappearance correlates with hearing losses. Outer hair cells are equipped with mechanotransducing stereocilia but in addition have a modified basolateral membrane which interconverts electrical to mechanical energy and so the cell can act as a force generator. The molecular basis of this interconversion is a motor molecule embedded at high density in the cell membrane which responds to electrical potential by undergoing a shape change. This class of cell appears to act as fast local motor element which provides local positive feedback and alters the mechanics of the basilar membrane motion. The net effect is that the local amplitude of the basilar membrane motion is enhanced by up to 100-fold.

Measurement scales

Reductionist explanations in science often contain, implicitly, the idea of a progressive shrinking of scale so that each stage in description concerns itself with progressively smaller physical dimensions. This principle has been used in the auditory system. At each level of description of hearing there is structure of functional importance. At the largest scales which need to be normally considered, the diffraction of sound by the outer ear colours the spectral components of the incoming sound and uses a description of sound on the scale of centimetres. On the other hand, the waves which propagate in the fluid of the inner ear have a wavelength of millimetres.

Structurally, the cochlea, the hearing organ of the inner ear, is characterized by a length scale of about 1 cm (10^{-2} m) (approximately independent of species). Thus all the mechanics of the ear take place on a scale of a centimetre and the description of wave propagation is described by a fluid mechanics which is non-turbulent. The associated fluid mechanics around the organ of Corti in the cochlea constitutes the macromechanics of the cochlea.

The next critical scale is 0.1 mm (10^{-4} m). This represents the scale of assemblies of cells which interact to modify the macromechanics of the cochlea. The organ of Corti is about 100 microns across and this scale represents the functional unit of the cochlea. Below that is a scale of the order of 1 μm (10^{-6} m), a dimension which defines the limit which can be seen readily in living tissue by light microscopy. This scale defines the mechano-electrical transduction processes on the hair cells and where the light microscope reveals organized structure. The mechanics of the cells at this scale comprise the micromechanics of the cochlea.

The scale of 10 nm (10^{-8} m) defines the scale of large molecules and assemblies of molecules out of which the transduction apparatus and the molecular motors of the hair cells are constructed. Finally, 0.1 nm (10^{-10} m) represents the scale at which physics meets biology and describes a scale where intramolecular events occur. This limiting dimension is needed in comprehensive descriptions of the movement of the basilar membrane as it is the scale which defines the mechanical perturbations at the threshold of hearing. In the original descriptions of the cochlea by von Bekesy, where measurements were carried out on 'dead' cochleas, the absence of physiological amplification meant that the movement of the basilar membrane extrapolated to threshold of hearing was estimated to be even smaller, about 10 pm (i.e. 10^{-11} m) (von Bekesy 1960). This figure was revised upwards in the early 1980s when the idea of cochlear amplification received considerable support.

Molecules

There has been an enormous growth in the application of molecular techniques to the study of hearing over the last five years, and most neurobiologists would anticipate that there should be rapid progress towards a cell-by-cell specification of the cochlea and the auditory pathway in the next decade. In part this is because techniques (such as amplification by the polymerase chain reaction) are now better matched to the extremely small quantity of material available in the inner ear. It is also because there has been a convergence of techniques from other areas of cell biology. There have also been better systems for studying hearing and some noticeable firsts have been scored using an interplay between mouse and human genetics. There are currently approximately 40 identified genes associated with deafness, some of which have a well described phenotype.

An interesting example of a congenital form of deafness is Usher Type 1B syndrome. This is a form of deafness that is often progressive and is often associated with malformations of the kidney. By the strategy of positional cloning this syndrome has been associated with the absence of a specific and known protein, myosin VIIA. As yet there is no convincing functional role for this myosin although other myosins are known to be involved in the transduction step in hair cells (myosin Ib) (Gibson et al 1995). In developmental studies, there is a particular interest in identifying cochlea-specific genes: one of these, Brn-3.1, has been localized specifically to the hair cells of the cochlea (Erkman et al 1996). At the moment, several genes of interest have been identified but there are considerable uncertainties about their precise functional role in the cochlea.

Putting it all back together

The cochlea therefore has a great deal of structure, all of which appeals to those with an interest in mechanics. The result of considerable research effort means that at the cellular level there is now a wealth of detail about the mechanisms of sound transduction by individual cells. Ionic channels in hair cells determine the codes sent to the brain by hair cells, but the question remains whether we have uncovered enough about the cells and their molecular components even to get us to the starting point of a realistic synthesis.

Cochlear feedback using both inner and outer hair cells involves multiple cellular elements. Would a sufficiently powerful computer be able to model the input–output properties of the cochlea? Some steps have obviously been made in this direction as the cochlea has been the target of numerous modelling attempts since the early 1940s. Even so, there are disagreements about what can be modelled. The main problem is to describe the response of cochlear mechanics to time-varying sound inputs, the essence of a speech processing mechanism. Technically this problem can be solved relatively easily in the frequency domain where the solutions are steady state. The difficulty with the time domain solution is that the contributions of interacting assemblies of around 1000 cells have to be calculated. Most analytically soluble approximations tend to throw the interesting physiology out at the same time. It is certainly true that time domain syntheses are possible but they are sensitive to the assumptions put into the model (as an example of a recent attempt to do this, see Nobili & Mammano 1996). So far most computer simulations have been also costly in computing time because the synthesis of the cellular information requires computing processes which are better carried out in parallel rather than sequential machines. Perhaps more promising here are electrical hardware models of the cochlea which can be regarded as real-time computers. Capturing the essence of the cell physiology has proved hard, and simply substituting silicon for cells has resulted in cochlear simulations which act as filter banks but do not have the same flavour of a real cochlear processor. It is worth stressing that the issues involved here are quantitative ones: simply providing a model which can select one sound frequency in an input stream is not enough: what any model must do is what real cochleas do, and that is to select between 3000 frequencies.

Hearing and the central nervous system

So far I have been discussing the cochlea because it provides the preprocessing stage for all subsequent analysis of incoming sounds. It therefore delivers tightly controlled frequency and timing information to the central nervous system. The pathways leading to the central nervous system are characterized by considerable afferent and efferent signal traffic. The anatomical structure alone shows that we can expect the description of sound processing to be complex and that the jump from neurophysiology to the psychophysics and neuropsychology of hearing is large. What we come to know about one species cannot necessarily be generalized to other species because of species-dependent differences in the pathways from the cochlea to the central nervous system. It thus seems likely that every species orders its auditory system slightly differently and that this in turn is related to the biological significance of the sound for it.

Are there general phenomena which cannot be answered by a reducing the auditory system to its component parts? Consider just two: one where simple modelling suggest that we can and one where, for the moment at least, we cannot and the reductionist programme has been less successful. The first example is the so-called Tartini (or combination) tone. If two tones of frequency f_1 and f_2 are presented to the subject (with $f_2 > f_1$), a 'virtual' tone of frequency $2\ f_2 – f_1$ can be heard. That this combination can be heard at all is an indication that the auditory system is not behaving as a linear system. There are therefore a number of possible sites where such non-linear signal transfer could be occurring. These include the synapses within the auditory pathways, the limited capacity of neural signal channels and mechanical non-linearities of the cochlea or even the mode of transmission of sound into it. However, the most likely source of the non-linearity is the mechano-transducing step in the hair cells (Jaramillo et al 1993). It is known the that this non-linearity is present as part of the signal in auditory nerve fibres and the mechanism is therefore a consequence of the mode of operation of the organ of Corti. This represents an example of a very obvious psychophysical phenomenon which can be traced to the molecular machinery of hearing.

An example where resynthesis of auditory function from a knowledge of the parts has been less successful is in the phenomenon of tinnitus or 'ringing in the ears'. This is a medical question for it has been estimated that tinnitus affects 15% of the population (although for only a small fraction of these does it represent a serious health problem). The origin of tinnitus is probably not the malfunctioning of the cochlea, despite a design in the cochlea which appears to be prone to spontaneous self-oscillation. The most likely explanation, by a process of elimination, is that it arises from central readjustment to an altered cochlear signal (in particular to the absence of the normal signals which are generated by the cochlea). The difficulty in providing a satisfactory explanation lies partly in knowing how to describe the phenomenon (do we have the correct language to describe the sensory attributes of tinnitus?), partly in developing objective measures of tinnitus (how do we investigate subjective sensation?) and partly in agreeing on an appropriate level of description (can it be reduced to describing neuronal interactions and, if so, which populations?). It seems likely that, given time, a satisfactory explanation will be devised, but for the moment we have here a definite limit to biological reductionism in hearing.

Acknowledgements

The work in my laboratory is supported by the Wellcome Trust, the Medical Research Council, the Royal Society and Defeating Deafness.

References

Ashmore JF, Kolston PJ 1995 Active processes in the cochlea. Curr Opin Neurobiol 4:503–508
Dallos P 1996 The cochlea. Springer-Verlag, New York
Erkman L, McEvilly RJ, Luo L et al 1996 Role of transcription factors Brn-3.1 and Brn-3.2 in auditory and visual system development. Nature 381:603–606
Gibson F, Walsh J, Mburu P et al 1995 A type VII myosin encoded by the mouse deafness gene *shaker-1*. Nature 374:62–64
Hudspeth AJ 1989 How the ear's works work. Nature 341:397–404
Jaramillo F, Markin VS, Hudspeth AJ 1993 Auditory illusions and the single hair cell. Nature 364:527–529
Nobili R, Mammano F 1996 Biophysics of the cochlea. II: stationary nonlinear phenomenology. J Acoust Soc Am 99:2244–2255
von Bekesy G 1960 Experiments in hearing. McGraw-Hill, Berkshire

DISCUSSION

Noble: What intrigued me about your discussion of levels is that you went from the top to the bottom, but you didn't say anything about the reverse direction. If we could, as it were, listen in to what's going on at those bottom levels, we wouldn't be hearing. That is a funny way to put it, but I hope it is clear what I am trying to say.

Ashmore: I absolutely agree with you: the extraction of critical speech information requires the interaction of a few thousand cells at a time.

Rose: In one case you're talking about molecules and structures, and right up the top when you're in psychophysics you're talking about processes: that is a different way of thinking about phenomena from thinking in terms of molecules and structures. So you can't simply, as it were, have an arrow from psychophysics to the CNS, because it's a different order of phenomenon.

Gray: Your process is one at the level of the CNS. You might very well by the standard processes of experimentation complete the story that goes all the way from electrobiophysics to auditory CNS, although you would still have the problems that we keep on revisiting as to what is the right level of discourse: is it the level of the system or the level of the molecule? Those arguments, I think, are not deeply mysterious or controversial. But when you get to the jump from auditory CNS to psychophysics, it's quite a different explanatory gap because at the moment we have no idea how any of this gives rise to conscious experience.

Quinn: Those gaps are somewhat less wide in studies of auditory processing in bats (by Suga's lab) and barn owls (by Konishi's and Knudsen's labs).

Raff: If you remove bits of the auditory pathway at various levels, starting at the highest level, how far can you go and still hear anything?

Morgan: The short answer is that I don't know, but there is a big auditory input to the midbrain. Presumably you could do something with that.

Garcia-Bellido: At which level do cochlear implants input information to the brain?

Ashmore: The cochlea implant programme is a technological spin-off of this sort of physiology. It attempts to provide a baseline of hearing for people who are essentially sensorineurally deaf: individuals in whom the cochlea is impaired, but the nerve itself is intact. Cochlear implants work by stimulating the intact nerve endings.

Garcia-Bellido: The nerves presumably carry different information depending on the level we are talking about. The sound when it comes to the nerve is not as specialized as its perception: it is just bringing frequencies to all the nerves. How does the brain discriminate different frequencies passing through to the nerve?

Ashmore: Cochlea implants don't restore perfect hearing, and therefore a major aspect of any individual being fitted with a cochlea implant is an extensive period of retraining the individual to use the limited information provided by the implant.

Morgan: What may be missing here is the notion that one of the ways the auditory nerve can carry information is by the frequency of neural firing, irrespective of which fibres are being stimulated.

Garcia-Bellido: How does that map frequencies to the brain? The mapping to the brain would be corresponding to some scalar value depending on which nerve you are dealing with.

Morgan: The point is that frequency information doesn't depend on which nerve is carrying an impulse, but rather on temporal phase locking of neural impulses. There is work that suggests that cochlea implants work particularly well combined with visual information. Isn't it the case that these implants are not much use for understanding speech unless you can see the person speaking?

Ashmore: Vision provides additional cues to hearing. The other feature of cochlea implants is that they are most effective in people who have become deaf after learning how to speak.

Nurse: Do deaf people who lip-read 'hear'?

Gray: A recent neuroimaging study using functional magnetic resonance imaging (fMRI) showed that silent lip-reading activated the same region in the auditory cortex that is normally activated in hearing people by speech (Calvert et al 1997).

Nurse: How is that established?

Gray: I can try to answer that by describing a parallel study. We are currently studying a group of synaesthetes who, when they hear words, see colours. They have done this all their lives and you can show that they're not faking: if you ask them what colour they experience with a particular word and you test them again without forewarning a year later, you get identical answers. If you ask non-synaesthetes to do the same thing, they don't come up with the same answers that they allocated a year before. The question arises: what is it that maintains colour experiences in such individuals when they hear words? There's absolutely no normal pathway of environmental reinforcement for that experience: nobody else sees the colours that they see when they hear words, and indeed synaesthetes keep this ability secret because they think themselves a bit peculiar. If you give these people a PET scan the same phenomenon occurs as in the silent lip-reading case: when these individuals hear words, there is increased blood flow in that region of the brain which normally deals with perception in the visual system. How and why this happens is unclear. The reason we're doing our study is to ask whether it is simply a consequence of associative learning or whether it is hard-wired. Our approach is also to study normal individuals whom we are training in the same associations that the synaesthetes spontaneously report. We are using fMRI in order to see whether, after training, these subjects show the same pattern of activation in the visual systems as do synaesthetes (Gray et al 1997).

Noble: The interesting thing about this is our reactions to it: we find it amazing. The point I want to make is that all higher order processing in the brain is amazing. It is our familiarity that is actually giving us the feeling there isn't a problem.

References

Calvert GA, Bullmore ET, Brammer MJ et al 1997 Activation of auditory cortex during silent lipreading. Science 276:593–596

Gray JA, Williams SCR, Nunn J, Baron-Cohen S 1997 Possible implications of synaesthesia for the hard question of consciousness. In: Baron-Cohen S, Harrison JE (eds) Synaesthesia: classic and contemporary readings. Blackwell, Oxford, p 173–181

The nested networks of brains and minds

Horace Barlow

Physiological Laboratory, University of Cambridge, Cambridge CB2 3EG, UK

Abstract. The reductionist approach to the brain shows promise of revolutionizing our ideas about what single neurons can do. A spine on a cortical pyramidal cell is about the size of a single *Escherichia coli*, and if the internal machinery of a spine is anything like as well organized as that of *E. coli*, the whole pyramidal cell with its 5000 spines must be capable of computations an order of magnitude more complex than those demanded of the neurons used for current models of the brain. These computations might enable single neurons to detect spatiotemporal patterns, i.e. Hebb's 'phase sequences'. Reductionism is apparently limited because its drive is to look for explanations at lower levels in the organizational tree. For this purpose it often uses isolated preparations in which such lower levels can be studied but higher levels cannot, because they have been thrown down the sink. Reductionism will never lead us to understand organization and interaction in parts discarded or ignored, and this must include the interactions between individual human minds that are crucial for understanding human society. Our brains possess a 'commentary system', a mechanism that can make reports on the internal status of some parts of the brain. This makes possible networks of minds, and the present meeting is such a network whose interactions are being recorded for posterity. On a grander scale such networking creates a cultural forum where communal goals and purposes are formulated, disseminated, modified, and often perpetuated in lasting form. The resulting group behaviour has obvious survival value, and is perhaps the feature that distinguishes humans most clearly from other species.

1998 The limits of reductionism in biology. Wiley, Chichester (Novartis Foundation Symposium 213) p 142–159

I was asked to talk about neural networks and reductionism, but although artificial neural networks have demonstrated the potentialities of a new style of computation, they have had very little success in what they were originally supposed to be doing — modelling the way psychological behaviour is produced through the interactions of real nerve cells. They typically have a miniscule number of elements compared with the real thing, the elements themselves are far too simple, and they are very richly and indiscriminately interconnected compared with the sparse and selective interconnection typically found naturally. There are good reasons for the way artificial neural networks were developed, but the final product is a bad model of a real brain. One of the positive results of the neural network approach is, however, very relevant to this meeting: it has been applied to the study of intracellular biochemical interactions, and this makes it clear that neurons can probably do much more than current models demand of them. We can hope that networks using elements that realize some of these potentialities will be much better models of the way the brain produces psychological behaviour, but at the moment we can only see this fascinating new generation of neural network models on the horizon.

As well as these new intracellular biochemical nets and the traditional interneuronal nets, there is a third level at which network-like interactions are important. The interactions of individual brains form, disseminate and perpetuate human societies, and this is the subject of the second half of my paper. Perhaps there are more than these three nested networks, for cells have nuclei as well as cytoplasm, and not all communication in human societies takes place directly between individuals, so human biology may require the study of network-like interactions at many levels. But although reductionism is often thought to direct attention exclusively to the molecular level, it will be argued in the final section that that this results from a misunderstanding of its nature, and that it need not be so limited.

A recent paradigm shift in our view of the mind

The reductionist approach to the brain has had an enormous impact. When I was starting to do research on vision 50 years ago, it was thought highly presumptuous to suggest a physiological explanation for any kind of psychological or mental phenomenon; these were supposed to have a non-material quality that could not possibly be explained by physical or biophysical processes. But now I think most scientists who are concerned with the problem take it for granted that the brain operates on physical principles and expect physiological explanations for its workings. There has, in fact, been a 'paradigm shift'; what we take for granted now is quite different from 50 years ago.

This is partly because we see that computers can do some, at least, of the things that previously required human minds, but it is also due to the outstanding success of reductionism in telling us about neurons, and we must first look where this is leading us.

New versus old neurons

Fifty years ago many thought that, if the brain was a computer, it was made of elements that did the sort of operation that Pitts & McCulloch (1947) proposed: neurons summed their excitatory and inhibitory inputs, and if the result was above a threshold value they gave an output impulse. It is easy to imagine such neurons performing logical AND, OR and NOT operations, so in principle they could realize any Boolean logical function.

It is now clear that the brain does not perform its computations in the way that that model suggested. Directional selectivity in the retina is probably the simplest well-studied form of logical computation, and a model proposed 30 years ago (Barlow & Levick 1965) is still probably the best representation of this logic. Responses to objects moving in the non-preferred direction are prevented by an inhibitory 'veto', rather than motion in the preferred direction being picked out by an AND operation, but there is a more important difference. In the original version the elements were identified with separate cells in the retina — receptors, horizontal cells and bipolars — and they were supposed to interact through conventional synapses. But anatomical evidence soon showed this must be wrong, and a newer version (Barlow 1996) differs in two radical ways: first the logical elements are not whole cells but sub-compartments of cells, and second an intracellular mechanism, calcium-induced calcium release, is postulated for the key operation vetoing responses in the non-preferred direction. There are reasons for believing that this is still not correct, but it demonstrates a new style of model in which the logical element has changed from a whole neuron to a sub-compartment of a neuron, and the proposed mechanism is intracellular rather than mediated by conventional synapses.

Our ideas about what can go on in such sub-compartments has changed enormously. Dennis Bray (1995) has worked out how the various components that control the flagellar motor of *Escherichia coli* interact using methods developed for artificial neural networks 10 years ago (Rumelhart & McClelland 1986). Those who contributed to the connectionist movement aimed at modelling how whole nerve cells interact to produce behaviour, but their work is now valued for having started a new style of parallel computation. Bray's use of these methods demonstrates that we have processes inside one *E. coli* that do what we used to think required many complete nerve cells. *E. coli* is about the size of a single spine of a pyramidal cell, and each pyramidal cell has about 5000 spines; its total volume is equivalent to about a quarter of a million *E. coli*, so these facts should bring about a revolution in the way we think about computation in the brain.

It now seems that the Pitts/McCulloch type of computation is only the tip of the iceberg; there is a whole tier of computation going on at an intracellular level that we have so far almost completely ignored. Table 1 shows crude estimates of the total power available for computations using a cell's membrane potential, and using processes in the cytoplasm beneath all its synapses. The comparison is made in terms of the number of nearly independent elements per neuron per second. The figures are extremely rough, but it is clear that the iceberg analogy is

TABLE 1 How many computational elements?

	Approx. time constant (s)	*Independent compartments per neuron*	*Elements per (neuron × s)*
Membrane potential	10^{-3}	<10	$<10^4$
Subsynaptic cytoplasm	10^{-2}	10^4	10^6

conservative; intracellular computation is slower, but much finer-grained than membrane potential computation, and may be 100 times as powerful.

It is important to realize that this new tier of computation is nested within the neurons, each of which contains separate intraneural machinery. Hence the input and output still have to pass through the communication channels to and from the neuron that we already know about, but the relation between input and output of a neuron can be much more elaborate than we previously supposed.

The main questions this raises are of course experimental: Can we find signs of this increased computational power in the relations between input and output of the neurons we actually record from? But we also need to think about the question theoretically: What could this additional computational power possibly be used for? And I think there is an obvious answer.

What could intraneural computation be used for?

Neurons are usually supposed either to have no memory (as in the Pitts/McCulloch neuron), or to have a long-term memory in the form of Hebbian synapses whose strength is modified only slowly. There have been proposals for a more rapid modification of synaptic efficacy (von der Malsburg 1981), but the enormous amount of computational power available prompts one to look for more radical possibilities than he proposed.

Neurophysiology has always lacked adequate explanations for the temporal aspects of perception and memory, yet these are obviously of crucial importance. Can we imagine that intracellular computational mechanisms fill this gap and make individual neurons the masters of timing in the brain? There is a natural tendency to think of memory as a separate organ in the brain because man-made memories use special technology and are spatially segregated from the processor, but perhaps each neuron has a memory of the temporal pattern of input at each of its synapses, at least over the preceding few seconds. Information about past inputs is potentially available to each neuron, and if it was stored and accessible for a short period a lot of brain-like computations could be done much more easily.

In a normal computer successive values of the input are stored at successive memory locations in a blind, unselective way, and the processor requires a vast addressing range in order to pick out the precise items it needs to operate on, since these are often not adjacent to each other. It is not immediately obvious what is equivalent to this addressing range in biological computation, but we know no way in which a neuron could selectively 'pay attention' to one of its inputs or to a subset of them, and even if it could a cortical pyramidal cell contacts a mere 10^4 input cells, a minute fraction of the range implied by an address of 32 bits or more. In parallel computation of the type used in artificial neural networks, and probably also the real brain, the processors have simultaneous access to rather small ranges of inputs, but all of them have a high probability of being relevant to the problem that has to be solved. In dynamic real life situations, however, recent past values of the inputs as well as their current values are likely to be biologically important, and what a parallel processor could compute would be dramatically improved if it had access to these past values. Figure 1 suggests that neurons might store successive values of each of their inputs for a limited period, which would give them immediate access to spatiotemporal patterns in their inputs, rather than just instantaneous, static patterns.

Donald Hebb, who invented 'Hebbian synapses' and 'neural assemblies', was responsible for starting people thinking in terms of neural networks (Hebb 1949). He called spatiotemporal patterns 'phase sequences' and he thought the ability of the nervous system to detect and generate them was of immense importance because it could account for key items of psychological behaviour, such as the ability to recognize the sequences of sounds in language, to play the piano, or to analyse movement in the visual field. His idea was that it was achieved through reverberating chains of activity in his neural assemblies, but even though we still do not know the biophysical mechanism for detecting any phase sequences, even directional selectivity, it now seems unlikely to depend upon reverberating networks.

A Hebbian synapse is one whose effectiveness is increased when its activation is followed by firing of the post-synaptic neuron and decreased when it is not. One can imagine mechanisms for Hebbian 'super synapses' (Barlow 1996) that store the average time interval between pre- and post-synaptic firing as well as the average effectiveness of the synapse in firing the cell. This could be achieved if there were several different mechanisms generating post-synaptic potentials having different time courses, with selective reinforcement according to when the post-synaptic cell fired. Though nothing of this sort has yet been found, there is recent evidence that the exact relative timing of pre- and post-synaptic activity has profound effects on the modification of synapses (Markram et al 1997).

Adding on-board memory for the recent past to the elements of neural networks would certainly enable them to do much more complicated tasks, and although

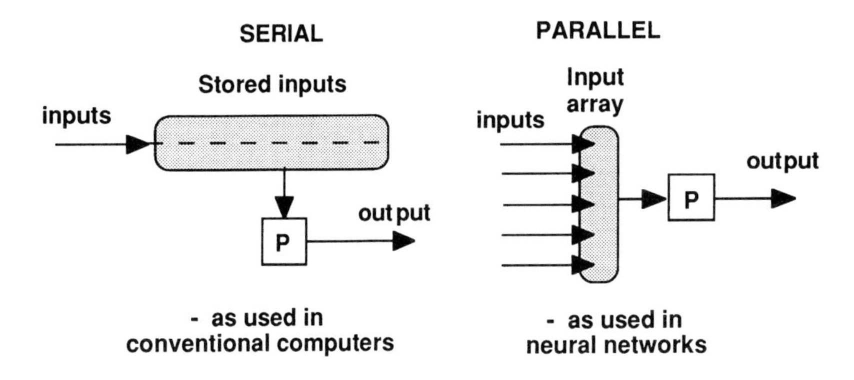

COMBINED SERIAL AND PARALLEL

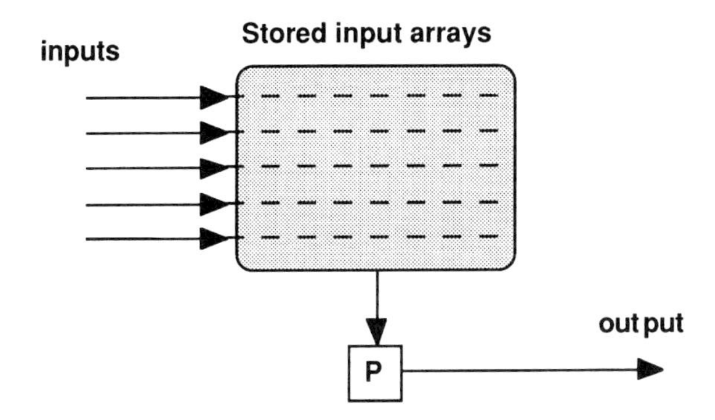

- desirable for networks handling sequences and rhythms

FIG. 1. Inputs to three types of computer processor. Diagram showing the input values that different types of computer processor can access directly. Whereas traditional computers must store all input values to which their serial processor may require access, a parallel processor has simultaneous access to a number of unstored inputs. It is argued here that neurons in the brain may have access, not only to many parallel inputs, but also to their values over the recent past. This would make them suitable for detecting Hebb's spatiotemporal 'phase sequences': with an auditory input they could differentiate speech sounds and melodies.

suggesting it is partly fantasy, when confronted with new computational mechanisms that are apparently unemployed there is no harm pointing out much-needed work for them to do — and the jobs for them to do really are ubiquitous: think of what is needed to distinguish speech sounds or to recognize a melody. There is hope that artificial networks employing such elements will provide more realistic models of real brains, but such developments lie in the future and will not overcome some apparent limitations of the reductionist approach.

Apparent limitations of reductionism

First notice something that is missing from both dualism and reductionism: neither of them deals happily with the intellectual products of our minds that fill libraries and museums. Karl Popper (1972) referred to these as the 'third world', neither wholly mental nor physical but partaking of both, and Merlin Donald (1991) attached great importance to this 'external memory' in the evolution of the mind. Even the purest reductionists among us must admit that these products of the mind are objective, often physical, and have had important material consequences for the way we live. Any paradigm for studying the higher functions of the brain must have room for them.

There is a *reductio ad absurdum* argument driving home the point that molecular reductionism is not enough. Jerry Lettvin (1995) once suggested that there might be 'mother cells' in the brain that were active when mother was seen, and not at any other time. Suppose for a moment that is true, and that we succeed in isolating them in full working order in a Petri dish. If we excite them electrically we can make the pattern of their responses just the same as those of similar cells in an intact animal seeing its mother, and their extracellular and intracellular processes, which are all that the reductionists can tell us about, would then be identical: are we to suppose that the subjective experience of 'mother' occurs in that Petri dish?

This dilemma has been in my mind for 50 years, ever since I recorded from retinal ganglion cells in the frog and was impressed that their responses had more similarity to my subjective visual experiences than the physical stimuli I was delivering to the retina. But I think that the idea that the contents of a Petri dish can have a subjective experience is absurd. The same biophysics may be going on, but it has none of the normal connotations of conscious experience.

Notice that whatever notion you have about the correlate of subjective experience in a cell or group of cells, be it oscillations, quantum coherence, or any other magic tingle, the possibility of recreating a subjective experience in a Petri dish is equally absurd. Subjective experience is an attribute of minds, not isolated nerve cells.

This is a useful thought experiment to illustrate a limitation of isolated preparations, which have been so important for the success of the reductionist approach. It is obvious that you cannot find out about what you have thrown away or deliberately chosen to ignore, and the usefulness of isolated preparations tends to make reductionists blind to any organization or interaction occurring outside such preparations. No wonder reductionists do not want to talk about minds, whose functional role in evolution certainly involves interactions that cannot be observed in Petri dishes containing neurons.

Ordinary ideas about other minds

We use the word 'mind' to describe what controls people's behaviour. Minds have stable attributes that can be reliably described: 'She has a subtle mind'; 'He is indecisive', etc. Minds can be sharp, blunt, conciliatory, aggressive, and so forth, and one can talk about them usefully and with a good deal of agreement. I think a large part of our brains must be concerned with assessing and storing knowledge of other minds; it's such an important concept to us so that we cannot possibly do without it.

Now there is nothing mysterious about how knowledge of other minds is obtained: it is by direct observation of repeated characteristics of other people's words and actions in the outside world, aided by gossip of course. But knowing about our own minds is different.

Direct subjective experience — a wrong and a right question

We know our own minds by direct experience. I think this is what people most want to hear an explanation for, but very little that is said about it makes any sense whatever to me and I think the question, 'What *is* direct conscious experience?' is a futile one. On the other hand, one *can* say sensible things about what these direct experiences enable people to *do*.

Suppose that in addition to the connections in the brain enabling it to respond to sensory messages with appropriate muscular movements, there are also connections enabling it to make reports on its own status, these reports serving a different function from the direct responses. Weiskranz (1997) suggested something like this when he proposed that what was missing in his blindsight patients was the activity of a 'commentary system'. To make this idea more precise, consider what a computer would need in order to implement such a system.

Meta-reports and the commentary system

The computer would need to be set up so that, in response to the appropriate command, it could at any time put a message on its screen saying, for instance, that it was in Microsoft Word with the file 'Novartis talk 1997' loaded, and that the cursor was at page 7 line 22. This would be a trivial addition to what is already available, but one can imagine it being extended in various ways. For example you might ask it why the cursor is on line 22, and one could arrange for it to explain that the word just typed would not fit on line 21; on further queries it might even go step by step through the algorithm that determines when a line feed is entered in the text file. Such meta-reports could also be made available to other computers making authorized enquiries, and the computer might be enabled to make requests of other computers asking them for meta-reports explaining what they were doing.

My own subjective experience seems to give me the ability to report on the status of my brain in this sort of way, and although the analogy does not *explain* subjective experiences, it does meet the objective of indicating what subjective experience enables us to *do* that we would not otherwise be able to do.

A reductionist working on an isolated preparation cannot tell you about the bits he has thrown away; in the same way, if you do not recognize the fact that there are other individuals having minds that control their behaviour, and that these minds can influence yours and you can influence theirs, you cannot understand what subjective experience enables you to do that you could not do without it. You have to take other minds into account for your own experiencing mind to make any sense. On this view there would be no point in having subjective experiences without other minds to communicate them to, just as there would be no point in the computer being able to explain its actions if there was no one to make the queries and read the meta-reports.

A hundred years ago, Nietzsche (1887) wrote: 'Consciousness is really only a net of communication between human beings; it is only as such that it had to develop; a solitary human being who lived as a beast of prey would not have needed it'. The few hundred words from which this is quoted argue that the improved mutual communication that results from consciousness accounts for its evolutionary value. Some find this obvious, others declare it to be nonsense, but I find it gives enormous insight that dawns rather slowly and then illuminates the whole problem. The survival value of consciousness was the point that most puzzled William James, but who can doubt that a major factor in the development of human civilizations has been conscious communication between beings who treat each other as parts of a conscious network?

I think a new paradigm is needed. This would take for granted that there is no qualitative difference between conscious and non-conscious neuronal mechanisms, and it would not be at all antireductionist so it could build on all that the molecular reductionists tell us about the neuronal mechanisms they discover. However, it would also build on Nietzsche's insight about networks of communicating humans, and would therefore turn our attention to problems different from those that interest a molecular reductionist.

An example is illustrated in Fig. 2. It must be pointed out that the messages passing along the lines in this figure are enormously more complicated than

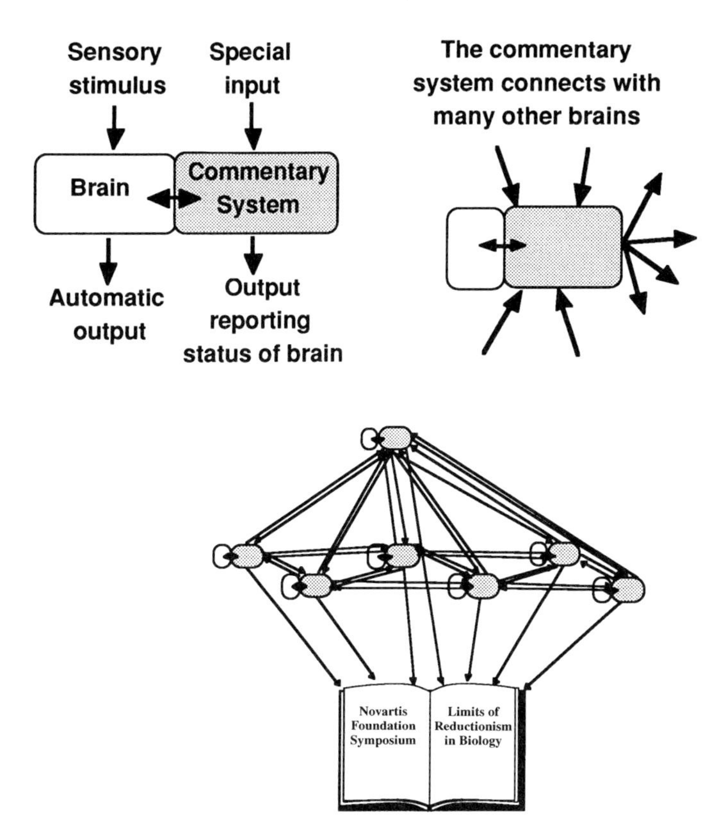

FIG. 2. Nietzschean networks of brains and minds. Brains having the 'commentary system' that Weiskranz (1997) proposed can respond to queries from other brains with reports on some aspects of their own internal status. This makes possible networks of many interconnected brains having properties reminiscent of traditional neural networks. These are the conscious networks that Nietzsche (1887) envisaged, and with the appropriate genetically determined value systems they could construct Popper's 'third world' and Donald's 'external memory'. The lower part of the figure illustrates the role of such a network in producing this volume.

those in the typical neural network figure, or in Bray's diagrams analysing intracellular biochemical networks. Nevertheless, social interactions occur in nets of communicating individuals, and they influence subsequent interactions along these connections; in that respect they are like classical connectionist networks.

Conscious networks can promote group survival

Communication between brains would not necessarily be beneficial, but the following three steps should form a network with properties that would promote group survival.

(1) *An individual brain must be able to report its own status to other brains and to request and understand reports from other brains about their status. This makes possible the formation of a network of communicating brains.*
These are the conditions for the network to be set up. The reports themselves are transmitted by motor movements, usually speech in a language understood by the community, and they are received in the same way as other sensory information. But such a network might simply spend its time circulating purposeless messages interfering with members of the community getting on with useful jobs. Groups of humans sometimes do just this, so more is needed to make the network useful.

(2) *The brains in the network must have a shared value system that promotes use of the network to formulate and disseminate communal knowledge and communal goals; this requires an innate, genetically constructed mechanism.*
When one encounters an unexpected biological feature, whether it is an anatomical structure, a new protein molecule or an item of behaviour, an important question to ask is 'How has it promoted the survival of the organism so that the controlling genes are retained?' This second step suggests the answer for the network of consciousness: the advantage comes through modifying group behaviour. This raises the familiar problem of group selection, though it is no more severe than for any other group characteristic. Presumably it is mainly through the selection of individuals that the species genome is modified, and the individual must increase its inclusive fitness more through participating in the network of brains than it would through an alternative, more selfish-seeming behaviour. This conflict between individual and group is surely a reality of life.

(3) *The innately determined value system must also drive individuals to perpetuate the products of the network by storing them in libraries, institutions and museums, thereby providing the basis for Popper's third world and Donald's external memory.*
Neither dualism nor reductionism take much account of the products of the mind that are stored in libraries and museums, in spite of the fact that these are objective, physical things that have obviously had a huge influence on us. It may seem outrageous to postulate a genetic tendency for humans to perpetuate knowledge in this way, but if genes build the dog's brain in such a way that it buries bones, they can surely build a human brain so that it buries its valued intellectual products in books and libraries. Considering the

influence accumulated knowledge has on our lives, this is not the futile exercise that one might at first suppose.

What does reductionism reduce?

Figure 3 illustrates a common view of reductionism in which its drive is always to explain phenomena that occur at high levels of organization in terms of components and their properties found at a lower level. The apparent limitation of this approach is that the isolated preparations used often do not allow interactions at higher levels to be studied directly, and it is often not even asked why the arrangements at the lower level are as they are found to be. Thus, if there

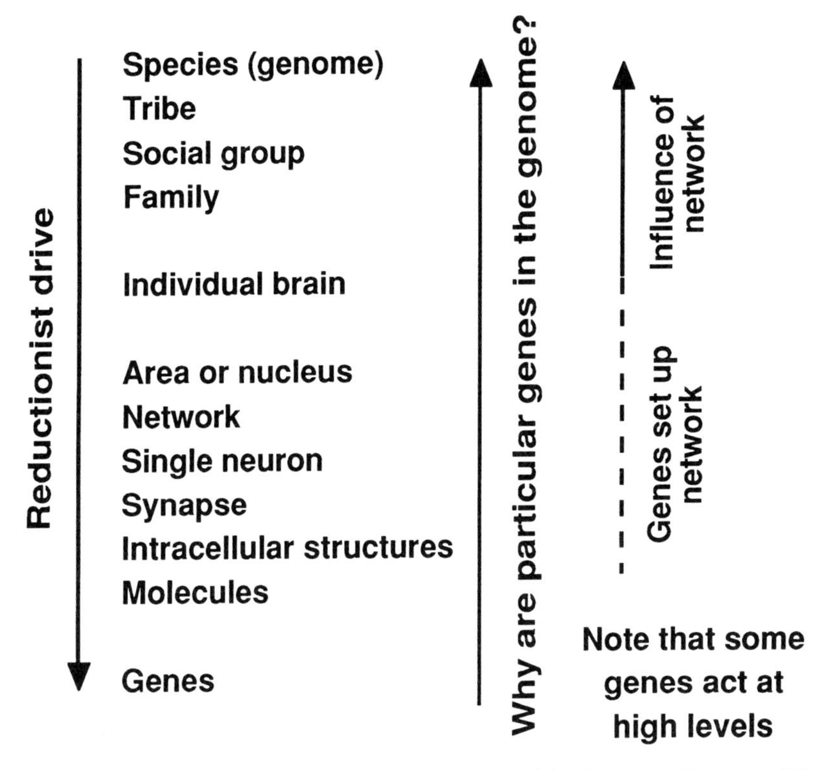

FIG. 3. Levels of organization. Reductionists like to explain phenomena in terms of simpler, more basic entities, ultimately genes, but to explain why particular genes are in the species genome one sometimes has to look at the organization and interactions of complex, high level entities such as the family and tribe. The neural networks that genes construct would not operate in many of the isolated preparations reductionists use.

are genes that build mechanisms setting up Nietzsche's 'networks of conscious communication', a molecular reductionist might find the genes, but would be most unlikely to find what they do. For this reason a molecular reductionist is certainly well-advised to cast an eye backwards over his shoulder at higher level phenomena, and this is obviously done; the clue to DNA's role did not come from its molecular structure but from its ability to change the pathogenicity of pneumococcus, and we all pay heed to evolution. But is this backward glance treacherous and antireductionist, or is it a valid part of reductionism?

Here I am going to appeal to an argument that stems from the ideas of Mach (1886) and Pearson (1892) about 'Economy of thought': it is that scientific explanation, and hence reductionist explanation, is driven by economy, for it aims to reduce the total length of the description required for a particular phenomenon (Barlow 1959, 1974, Watanabe 1960). We know a lot and are finding out more about the chemical constituents of living things; furthermore this knowledge seems to be permanent, and not to require substantial modification with each new generation of investigators. By building on this knowledge we can, as it were, reduce the surprise value of new biological phenomena as they are discovered, and thus reduce the length of the additional description required for them. In information theoretic terms, we are using known redundancy (i.e. known patterns and regularities) to recode more compactly the messages we are receiving — messages that without such recoding would occupy much more of the limited channel space that our minds are able comprehend. The reductionist drive is often as it is shown in Fig. 3, but this is because the reliable knowledge we can build on is at the molecular level. This is the closest we can get to explaining things in terms of 'universal principles governing ... common ultimate constituents', which is the definition of reductionism given by Thomas Nagel (1998, this volume). Reductionism as the heuristic of economical description does not conflict with this, but allows us to use principles discovered at non-molecular levels.

Conclusions

Civilized communities are networks of nested networks. The interactions of individual brains with each other and with history make human communities what they are. Each brain is a network of interacting neurons, and each neuron is a network of interacting molecules. Our knowledge is more reliable and more complete at the molecular level, and for that reason the reductionist drive towards explaining biology in terms of molecules has been extremely successful, but knowledge of organization and interaction in the outer layers of the nested networks may be required to understand the inner layers, as well as vice versa.

Reductionism makes biological phenomena more intelligible by reducing the length of the descriptions they require, not necessarily by reducing everything to molecules. On this view, all good science is reductionist as Lewis Wolpert proclaims (in his introduction to this volume), and limitations arise only from misunderstanding the nature of scientific explanation.

References

Barlow HB 1959 Sensory mechanisms, the reduction of redundancy, and intelligence. In: The mechanisation of thought processes. HMSO, London, p 535–539
Barlow HB 1974 Inductive inference, coding, perception, and language. Perception 3:123–134
Barlow HB 1996 Intraneuronal information processing, directional selectivity and memory for spatio-temporal sequences. Network Comput Neural Syst 7:251–259
Barlow HB, Levick WR 1965 The mechanism of directionally selective units in the rabbit's retina. J Physiol 178:477–504
Bray D 1995 Protein molecules as computational elements in living cells. Nature 376:307–312
Donald M 1991 Origins of the modern mind: three stages in the evolution of culture and cognition. Harvard University Press, Cambridge
Hebb DO 1949 The organization of behaviour. Wiley, New York
Lettvin JY 1995 Appendix to: The neuron doctrine in perception. In: Gazzaniga M (ed) The cognitive neurosciences. MIT Press, Cambridge, MA
Mach E 1886 The analysis of sensations, and the relation of the physical to the psychical (Engl transl of the 5th German edition by S Waterlow) Open Court, Chicago and London (1959 reprint; Dover, New York)
Markram H, Lübke J, Frotscher M, Sakmann B 1997 Regulation of synaptic efficacy by coincidence of APs and EPSPs. Science 275:213–215
Nagel T 1998 Reductionism and antireductionism. In: The limits of reductionism in biology. Wiley, Chichester (Novartis Found Symp 213) p 3–14
Nietzsche F 1887 Die Fröhliche Wissenschaft. Verlag von E W Fritzsch, Leipzig (Engl transl by W Kaufmann 1974: The gay science. Vintage Books Random House, New York)
Pitts WH, McCulloch WS 1947 How we know universals: the perception of auditory and visual forms. Bull Math Biophys 9:127–147
Pearson K 1892 The grammar of science. Walter Scott, London
Popper K 1972 Objective knowledge. Clarendon Press, Oxford
Rumelhart DE, McClelland J 1986 Parallel distributed processing (three volumes). MIT Press, Cambridge, MA
von der Malsburg C 1981 The correlation theory of brain function (Internal Report No. 81-2). Department of Neurobiology, Max-Planck Institute for Biophysical Chemistry, D3400, Goettingen, Germany
Watanabe S 1960 Information-theoretical aspects of inductive and deductive inference. IBM J Res Dev 4:208–231
Weiskrantz L 1997 Consciousness lost and found. Oxford University Press, Oxford

DISCUSSION

Raff: Suppose you had mapped 5000 neurons in a human brain that fired with the experience of 'mother', and you had a way of keeping the head alive separate from the body. Could you imagine it working so that it could appreciate the experience of mother? Although the 5000 cells alone in a dish could not reproduce this experience, if some day you were to put all the billions of cells together in a dish, connected to one another just as they were in the brain, is there any intrinsic reason why the assembly of cells could not function as they did in the person?

The reason I ask this is because some think that there is more to higher brain function than the brain itself. But if you believe that there isn't anything more to conscious experience than cells and molecules working in complex ways, then someday, in principle, you could hope to understand it, and even study aspects of it in a dish.

Barlow: It's what consciousness *does*, not what it *is*, that matters. If it mediates social communication among humans through the commentary system, then its adequate study in a dish or an isolated brain is not possible, even in principle.

Gray: There is one piece of evidence relevant to this. In split brain preparations where a commisure is cut between the two halves of the brain, Gazzaniga (1998) showed that you have to attribute some form of consciousness to both the left and the right halves of the brain. So, if you started out with this system, which presumably like the rest of us has a unified consciousness (whatever that means), and you cut it in two, this raises the question of what happens if you now cut it in four, and then in eight and so on. We just don't actually know at what level it will get too small to sustain consciousness.

Still on the same point, Christof Koch and Francis Crick are seriously talking about looking for cells in the brain which fire when there is consciousness and then comparing them to cells which don't (Koch 1998). They intend to ask whether this firing subset of cells have specifically different genetic properties and so on. Although this seems highly implausible, it is not something we can rule out. If it's something that were to be found to actually happen, Martin Raff is right: you might be able to grow them in the Petri dish.

Rose: Martin Raff has asked what is essentially a Laplacian question. Laplace claimed that if you knew the positions of every particle in the universe at moment *a* you would able to predict the future of the universe at some point *z*. We know that he was wrong.

Wolpert: It is not the same question.

Rose: It is, because Martin Raff is asking whether, if you were to put all these cells together in a particular way, you would then be able to re-create the experience of mother. In a sense, I think the answer is 'yes' if you know one additional thing, which is what Horace Barlow's Petri dish didn't give us — the history of those cells.

Raff: I am saying that the history is already there. Nothing has changed. You simply take a head that has been 'programmed' with the experience of mother and put it in a dish.

Rose: If the brain had that past history, then I suspect that the answer actually has to be 'yes'.

I wanted to come back to a couple of things in Horace's paper. At the beginning you were saying that there are computations that are embedded within cells as well as taking place between cells. My suspicion is that the cell needs an awful lot of those computations simply to exist, before it begins to act on the external world. What the nervous system gives us in between-cell interactions is the capacity of interacting with the external world as well as merely existing.

Barlow: That's obviously right: a lot of intracellular computation may be just maintenance work. On the other hand, we know that *E. coli* does proper computations, and I don't see why cells in multicellular organisms shouldn't also be doing these internal computations. To assume that it must all be mere maintenance seems to me an extreme point of view.

Rose: I was interested that you resurrected Popper's 'third world', because I always thought that Eccles and Popper in that sense were 'troilists' rather than dualists, because they had this combination of three. In a sense, if you had replaced what you called a meta computer in your network model with a soul, you would have recreated Descartes!

Barlow: Of course, I can tie myself into knots by asking this question about 'what is consciousness?' You can do the same thing in vision with the experience of 'redness'. But the interesting aspects of colour vision are not really the nature of the individual's experience of colour, but what enables that individual to discriminate in the outside world: in other words, what colour vision enables a person to do. I'm urging that people should put this question about what is the actual conscious experience on the back burner and concentrate on what it enables people to do, because that after all is what's going to determine the survival value of this phenomenon.

Bateson: If there is a function of consciousness, then it must have been seen by natural selection — it must be observable in some sense. That then raises the question of what would be an effective research strategy to study the utility of consciousness. How would one actually show that one part of consciousness was needed for a particular aspect of a social transaction, for instance? I can see that there might be a way of doing this by using brain-injured people or drugs, but it seems to me that we have to do empirical work of this kind in order to show that a person who is unable to do something specific lacks some feature of consciousness.

Barlow: To study this we need an equivalent of colour blindness in the field of social communication. Then we could get a colony of these people to see how well they do! My prediction would be to that for a social animal you have got to have a social communication system, which seems to quite obviously depend upon consciousness. You can't hold a meeting such as this without each of us being conscious of what other people are saying. Autistic individuals seem to me the nearest we have to people who fail to communicate in a way that is the essential function of consciousness.

Noble: I was excited by your presentation, because it seemed to me that you got to the essence of one of the key issues facing us in this discussion. You said something along the lines of 'Dear me, I've almost convinced myself to be an antireductionist', and then you stepped back on two grounds, and I want to challenge both of them. You stepped back first because you said that there was much more explanatory power going from lower levels upwards than in the other direction. I'm not sure that is true, and this is why I said yesterday that genes are captives to the successful physiological systems that contain them. I could equally have said they are captives of the sociological systems in which we are. I don't think we've yet teased that out to see how much balance there is of explanatory power here between the bottom up and top down views, and I'm not myself totally convinced that it does all lie in saying that one of these arrows is very strong and the other one very weak; I think that's in part a cultural thing. If I may as it were state the manifesto for the antireductionists in this meeting, it is that culturally over a period of 30–40 years we have gone through an extremely strong period of reductionism in the way in which science is pursued. It's not surprising therefore (if I may be a social analyst of your performance for a moment) that you hesitate at the last minute before jumping over to the antireductionist camp! That's my challenge to you, and I would like you to justify your hesitation. But also in your presentation you said that we can't do without the concept of mind. It is of great importance to the discussion which I'm trying to precipitate here, to clarify whether that is a matter of convenience or necessity. A key feature of the antireductionist manifesto is that this is a matter of necessity: it is incoherent to deny mind. This is the centre of the whole question.

Nurse: I think this issue of whether discoveries at lower levels always explain higher levels and vice versa is key to this debate. I want to propose a thought experiment. Let's imagine that Crick and Watson existed before Mendel: in other words, we have the structure of DNA before we had the function. It would have been meaningless. It is only knowing what it does at the higher level — i.e. that it replicates and has to copy itself — that endows the structure with meaning. In other words, the higher level problem is illuminating the lower level.

Barlow: That is why I said that what you do get by looking at the higher levels is good questions, but you don't get the answers.

Nurse: I would say you get some meaning.

Barlow: Yes, I agree. As I explain at the end of my paper, one of the reasons that reductionism (in the sense of reducing biology to physics and molecules) works is because you then build your explanations on a secure foundation of reliable facts and theory. The result is that the *additional* explanation required for new biological phenomena is reduced: you can shorten the necessary description. You don't generally get this advantage when you try to base your explanations on sociology or accounts of biological organization found at higher levels. A possible exception is the conceptual simplification you get when appealing to evolution and natural selection.

References

Gazzaniga MS 1998 Evolutionary perspectives on conscious experience. In: Jasper HH, Descarries L, Castellucci VF, Rossignol S (eds) Consciousness. Lippincott-Raven, Philadelphia, in press

Koch C 1998 The neuroanatomy of visual consciousness. In: Jasper HH, Descarries L, Castellucci VF, Rossignol S (eds) Consciousness. Lippincott-Raven, Philadelphia, in press

Genes, environment and the development of behaviour

Patrick Bateson

Sub-Department of Animal Behaviour, University of Cambridge, Madingley, Cambridge CB3 8AA, UK

Abstract. Explanations of where our behaviour comes from are frequently presented in terms of the exclusive importance of one set of factors, either genetic or environmental. Unravelling the external and internal sources of individual differences is a useful first step in analysing behavioural development. Nevertheless, the analytical method that was well designed for extracting influences from a confusing mass of data was never a substitute for a theory. It was simply a means to an end. Descriptive statements about the genetic and environmental sources of variation in the population do not offer an adequate basis for understanding what happens to individuals. That awareness was an important step in moving towards an adequate theory of behavioural development. As an example of how that may be done, I discuss the interplay between the developing individual and its environment in highly regulated learning processes such as imprinting. Getting the level of explanation right is crucial. A purely molecular or synaptic account of the processes involved in the development of behaviour is inadequate. Nevertheless, those connectionist models that are properly rooted in a thorough knowledge of behaviour and physiology do provide a promising route out of the reductionism and the empty interactionism that characterized the old nature–nurture debates.

1998 The limits of reductionism in biology. Wiley, Chichester (Novartis Foundation Symposium 213) p 160–175

Animals show specific preferences for things without prior contact with what they prefer. They also exhibit highly organized motor patterns without practising what they perform. Should these aspects of behaviour be set in opposition to those that are changed by experience? Many have argued in the past that two different systems are involved in development and have gone on to suggest that one is required for innate behaviour and the other for acquired behaviour. The developmental processes giving rise to predispositions on the one hand and learning processes on the other were supposedly separable in terms of the forms of behaviour they produced. Modern writers are less likely than before to conflate process and outcome. Even so, a system of behaviour is commonly referred to as 'genetically determined' or simply as 'genetic'. Such thinking is apparent in the modern approaches to evolutionary psychology (e.g. Barkow et al 1992) and in ideas about domain-specific systems of knowledge (e.g. Carey & Spelke 1994).

Sources of difference and interactions between them

Robert Hinde (1969) was one among several powerful critics of behavioural dichotomies. He suggested, however, that recognizing two sources of *difference* did have promise as an analytical tool. This approach is still the mainstay of a great deal of human behavioural genetics. In the comparison of genetically identical twins with dizygotic twins, for instance, the aim is to attribute variation in behaviour to genes or to the environment.

The temptation to confuse the analytical process with explanation is great and implicitly the dichotomy of sources of variation has been used to justify the notion of an additive developmental process. Such notions immediately collapse in the face of evidence for statistical interactions between the sources of variation. Over the years more and more examples have been appearing from both plants and animals of what may be called the 'jukebox'-like character of development. The same genotype can produce astonishingly different phenotypes, depending on individual experience. The caste differences in social insects provide a striking example. Some normally green grasshoppers growing up on African savannah blackened by fire are also black and prefer black backgrounds (see Rowell 1971). As a result they are less easily detected by predators. However, their offspring, developing among new grass, suppress the mechanisms making black cuticle and are once again green. The adaptive significance is obvious but details of the mechanism remain to be worked out. However, cases are known where particular genes are only expressed in special environmental conditions, as in the temperature-sensitive mutants (e.g. Wu et al 1978).

The study of individual differences in behaviour has been revolutionized in recent years by the discovery of similar cases throughout the animal kingdom. More and more examples of striking differences in reproductive behaviour are being found between members of the same species which are of the same sex and age (Caro & Bateson 1986). Each individual is capable of developing in more than one way — a jukebox with the potential for playing many tunes but, in the course of its life, playing only one. The particular tune it does play is triggered by the conditions in which it grows up.

The analytical method that partitions variation into genetic and environmental components may be misleading in a different way. A failure to find a statistical interaction does not mean that the developmental process is additive. In studies of children who have been adopted, it is sometimes possible to compare their performances on a given scale of measurement with those of their adopting parents and their biological parents. In such studies it commonly seems to be the case that no statistical interaction is found between the scores of the two types of parents (see Plomin et al 1988). The effects of the genes, provided by the biological parents, and the effects of the environment, provided by the adopting parents, seem to add together in simple linear fashion. Together the biological and adopting parents' scores account for about 10–20% of the variation in the children's scores (e.g. Horn 1983), but there is no interaction.

To make the argument simple, let us assume that initially the cognitive abilities of the child are most strongly affected by their biological parents but that they are increasingly affected by social transactions with their adopting parents. The quality of the environment provided by the adopting parents may not merely depend on their own cognitive abilities but also on the difference between their characteristics and those of the child. Disappointed adopting parents might provide a much less supportive environment for a dull child than those whose expectations were fully satisfied by the responsiveness of a bright child. Conversely, a potentially bright child adopted by dull people might be less challenged and more frustrated than if he or she had been adopted by bright people. Here again the difference between the child and its adopting parents matters, but this time in the reverse direction. As a result of transactions with their adopting parents, the bigger the absolute difference between the biological and adopting parents, the more adversely affected will be the children. Robert Plomin found in re-analysis of data from the Colorado Adoption Study that this was, indeed, the case (see Bateson 1987). I explored this effect further in some computer simulations. When the average difference in cognitive ability between the biological and adopting parents is small, a significant statistical interaction is found, but when the average difference between true and adopting mothers is larger, the interaction disappears. However, the effect on the child of the difference between the biological and adopting parents gets more pronounced (Bateson 1989). Leaving aside all the simplifying assumptions in this story, a clear message emerges. Unless we go in armed with a specific hypothesis about the dynamics of what happens in the child's development, conventional statistical analysis is liable to give a completely misleading picture.

While the modern emphasis is on the interplay between the developing organism and the conditions in which it finds itself, the fashionable word 'interaction' is often reduced to meaning that statistical term which results from a non-linear relationship between independent variables. At best, analyses of statistical interaction should be the starting points of attempts to understand how developmental processes work and should not be treated as ends in themselves.

The dynamic picture that emerges from empirical studies of behavioural development is one that does not sit comfortably with static notions of two types of behaviour, innate and learned. Undoubtedly, debate has been hindered because of all the different meanings that have been attached to 'innate' or 'instinct' (see

TABLE 1 Various meanings of innateness and instinct

Present at birth
Not learned
Develops prefunctionally
Unchanged once developed
Shared by all members of the species
Formed by a distinct behavioural system
Served by a distinct neural module
Adapted during evolution
Behavioural differences due to genes

Table 1). These are not the defining characteristics of a single class of behaviour patterns since one of the connotations of innateness does not necessarily apply to a given pattern of behaviour when another has been demonstrated. If a behaviour pattern develops without obvious practice or example, the control of that activity and its form may subsequently be modified by learning. A good example is the development of the human smile.

The point is that, in thinking about innateness, people have been focusing on very different matters. The evolutionary and ontogenetic origins are in play at the same time as considerations of the outcome of the regulatory and dynamic aspects of development. We need to see how these various strands may be brought together. A good case on which to focus is imprinting.

Imprinting and neural nets

Imprinting is involved in a process by which animals form exclusive attachments to a care-giver or future sexual partner. This is a learning process, but before learning takes place, the animal has clear preferences for the type of stimuli which it will learn about. It also has in place a suite of responses which it will direct towards those stimuli. Learning takes place at a biologically appropriate time in the life-cycle and few would doubt that the whole process has been adapted during evolution for the kin-recognition function which it currently serves.

In precocial birds, such as domestic chicks, filial imprinting occurs shortly after hatching and may simply require exposure to a conspicuous object. Even in very young chicks, transfer of training occurs; animals that have been imprinted may be influenced by this experience when they are given a rewarded discrimination. The nature of the transfer depends on how they have been imprinted and what they are required to discriminate between. If they are asked to discriminate between two objects with which they have been imprinted, then they will benefit from imprinting if the exposure was separated in time. However, their performance will be impaired if exposure to the two objects was close together in time (Chantrey 1974, Honey et al 1993). The advantages of using the same information in a variety of contexts can be great. The mother is an ever-present figure in the young animal's life. Her actions may prove extremely important in predicting where it can find crucial resources for itself. While the mother gives signals such as the food call which the chicks respond to without learning, all the chicks hear this and they are in competition with each other. So the smartness involved in transfer of training will pay the individual possessing such attributes.

In one set of experiments, chicks were trained with a flashing, rotating light or with a rotating stuffed jungle fowl and then given a choice between them (Horn & McCabe 1984, Bolhuis et al 1985). The artificial and naturalistic stimuli were matched for their effectiveness in eliciting approach from day-old domestic chicks by varying the rate of rotation of the object. Despite the matching, the stuffed jungle fowl became more attractive than the box in untrained birds by the second day after hatching. The shift towards a stronger fowl bias was also apparent in birds that have been imprinted with either a fowl or a box. Features of the jungle fowl that make it especially attractive as the predisposition emerges are located around the head. However, they are not specific to jungle-fowl since a stuffed rotating gadwall duck and a polecat were equally attractive (Johnson & Horn 1988).

Gabriel Horn and I have attempted to produce a theoretical account of imprinting that helps us to bring together the predispositions of animals that imprint with the effects of experience on their preferences (Bateson & Horn 1994). The first step involves detection of features in a stimulus presented to a young bird. Aspects of the stimulus which the bird is predisposed to find attractive are thought to be picked out by analysis at this stage. The second step involves comparison between what has already been experienced and the current input. Of course, before imprinting has taken place, no comparison is involved. Once it has occurred, recognition of what is familiar and what is novel is crucial. Finally, the third stage involves control of the various motor patterns involved in executing filial behaviour. This model is given the acronym 'ARE' for analysis, recognition and execution (see Fig. 1).

In creating a neural net that conformed to this flow chart, we tried to ensure that most of the sub-processes are as plausible in neural terms as current knowledge allows and that the whole system has the behavioural structure of an intact animal (Bateson & Horn 1994). The model exhibits a well known feature of behavioural development seen in animals, tending to settle into familiar habits, while also able to build with increasing elaboration on the basis of previous perceptual experience. The model simulates the closure of the sensitive period in

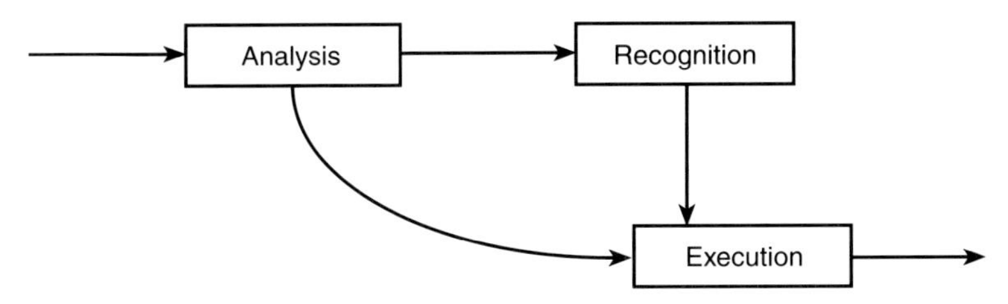

FIG. 1. Organizational model of the basic sub-processes involved in imprinting.

development, which is known to be dependent on experience, at least in part. The greater the activating value of the features in the experienced stimuli, the more quickly the sensitive period comes to an end. Above all, the various influences on behaviour do not add together. As in any non-linear system, small changes in certain parameters sometimes make big differences to the outcome and large changes in others have no effect whatsoever.

The case for some theory of mechanism is that it spawns explicit working models which bring with them mental discipline and expose weaknesses in a verbal argument that are all too easily missed. They can show how we are easily misled by the dynamics of development into supposing that the processes are so complicated that they are beyond comprehension. From the point of future empirical research, they can suggest profitable new lines of enquiry. They bring understanding of how real systems generate the seemingly elaborate things that we observe. Their predictions may be false, but they are worth testing just because the assumptions are rooted in psychological and biological reality.

An array of different neurobiological techniques have implicated the intermediate and medial part of the hyperstriatum ventrale (IMHV) on both sides of the brain as being sites of a neural representation of the imprinting object (Horn 1985, 1991). When evidence is open to a variety of interpretations, greater confidence in a particular explanation may be achieved by tackling the problem from a number of different angles. Each piece of evidence obtained by the different approaches may be ambiguous, but the ambiguities are different in each case. When the whole body of evidence is considered, therefore, much greater confidence may be placed on a particular meaning. An analogy is trying to locate on a map the position of a visible mountain top. One compass bearing is usually not enough. Two bearings from different angles provide a much better fix and three bearings give the most reliable position for the top. It was this approach that led to the conclusion that IMHV is a site of the neural representation of the imprinting object and correspond to the changed weights in the connection within the hypothetical recognition system of the neural net model.

Chicks that have had both left and right IMHV removed surgically are unable to imprint and if bilateral lesions are placed immediately after imprinting, the birds show no recognition of the imprinting object (see Horn 1985). However, it is possible for chicks to learn a rewarded discrimination after bilateral removal of IMHV. They will also learn to press a pedal rewarded by the view of an imprinting stimulus even though they do not go on to learn the characteristics of that stimulus (Johnson & Horn 1986). So it is possible to dissociate the neural mechanisms involved in imprinting from those involved in learning when an external reward is required. Nevertheless, transfer of training means that the two processes must be connected in the intact animal.

Lesion studies suggest that another representation (known as S′) is formed in another region of the brain about six hours after imprinting (see Horn 1985). This representation can be prevented by placing a lesion in the right IMHV soon after imprinting. The second store could provide the point of contact between imprinting and rewarded learning. We performed a behavioural experiment showing that when transfer of training is given immediately after imprinting, the rate of learning the discrimination is unaffected by imprinting. However, if discrimination learning was delayed by six hours then the type of pre-exposure did affect the rate of learning, with those imprinted with two objects close together in time learning significantly more slowly than the ones that were imprinted with two objects separated in time. In contrast, all birds showed a strong preference for the imprinting object immediately after imprinting. This experiment suggested that, while the memory required for recognition is formed quickly, the memory system sustaining transfer of training is not.

We showed that the functional characteristics of the two mechanisms differ (Honey 1995). Although both left IMHV and S′ supported imprinting preferences, only in the chicks with S′ did the memories formed during imprinting influence the acquisition of a heat-reinforced discrimination in which two imprinted objects served as stimuli.

I sketch in Fig. 2 a way in which the evidence may be brought together by building on the original ARE model. A second representation of the imprinting object is established at S′ and confirmed by a slow acting process originating in the recognition system involved in imprinting (R). This second representation can control pre-existing response mechanisms but may also control other motor systems as the result of external reward. Why are the two independent memory systems needed? Possibly because the associative process involved in external reward requires different rules from the simple recognition process involved in imprinting. Whether or not these speculations are correct, the main point of my story is to illustrate how knowledge is advanced by moving backwards and forwards between the behavioural and the neural levels. This stance differs sharply from a commonly held position that the people studying behaviour

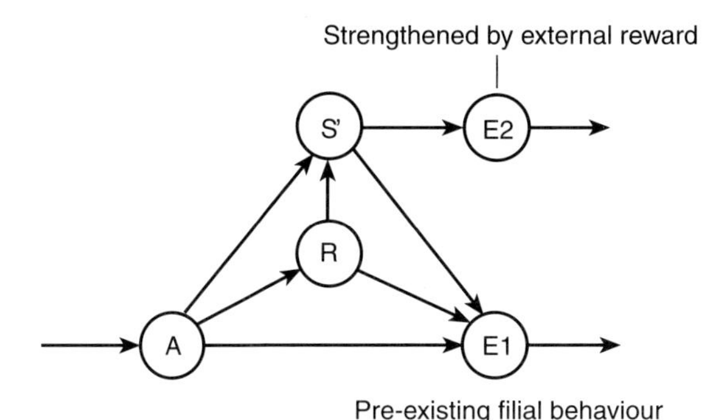

FIG. 2. Links between the sub-processes involved in imprinting and those involved in externally rewarded learning.

should pose problems for the neuroscientists who, having done their work, should doubtless pass on the project to the molecular biologists and eventually the physicists.

Different types of model

Some years ago, Alex Kacelnik and I planned a paper on the different ways in which people used models in behavioural biology. We suspected that we would be given a hard time by the philosophers of science and put the project on one side until we could do some more reading. Needless to say, the time never came, but I think some of the points we raised are relevant in the present discussion of reductionism. With Alex Kacelnik's permission, I want to mention our classification of models. It owes a lot to the three distinctions made by Marr (1982) between computational, algorithmic and hardware implementing approaches, but adds two categories. We have five classes:

(1) *Empirical models* reproduce observed relations between independent variables and observed behaviour without explanatory argument.
(2) *Functional models* relate the performance of a job to a given end-point such as survival or reproductive success.
(3) *Organizational models* set out the relations between different sub-systems involved in an overall process.
(4) *Logical models* give details of how processes might work by making explicit the logical steps or algorithms.
(5) *Mechanistic models* specify how rules in logical models might be implemented.

Four of these types of model could be seen as forming part of a coherent reductionist program. We start out with a precise description of the phenomenon and then we move to a flow chart of the processes involved. We begin to consider the rules that enable the processes to work and we finally start to consider how the rules might be implemented by plausible mechanisms. Work at each of these steps will be more efficient if the level is chosen with care. It could be argued that unifying explanatory principles for the organization of behaviour will never require detailed knowledge of what happens among the microtubules. Even so, the general approach that brings together behavioural biology and neurobiology is a programme with which the great majority working in this area would feel comfortable. In this chapter I have attempted to illustrate the process in my own collaborative work on imprinting.

One type of modelling lies at right angles to all the rest. That is the modelling involved in a functional approach which seeks to get to the specific way in which behaviour may maximize the chances of survival or maximize reproductive success. To give a non-biological example of what I mean by the 'specific way', it is obvious that the function of a corkscrew is not to get drunk but to get into the bottle that contains the liquid that enables us to get drunk. Similarly, the function of foraging is to find nutrients efficiently so that we may survive so that we may breed or help our relations to breed. In studying behavioural mechanism, the functional approach helps in framing questions and in noticing conditions that are likely to be important in controlling behaviour. But the process of investigating utility is not obviously reductionist in character.

Conclusion

Growing knowledge of what happens during neural and behavioural development should inhibit any residual dichotomizing tendencies. It should also make us deeply sceptical of the approach which merely seeks to parcel out variation into its likely causes without considering the assumptions that lie behind the method of parcelling or without asking what the developmental process might be like. In general, analysing sources of variation is a limited and sometimes misleading way to understand a process. Nevertheless, it remains true that underlying structure is crucial in the regulation of development. We see that most clearly in the jukebox-like character of some developmental processes. But it also obvious when we consider regulated processes of learning.

Behavioural biologists like me tend to say, 'Everything is so interactive that it couldn't possibly be reduced to what happens at a single neuron'. The behaviour of a free-running animal obviously depends on a large number of factors and the animal changes the conditions in which it lives as much as it is changed by them. When told about all the things required to generate an observed piece of behaviour, many neuroscientists react with irritation at what seems to them to be a blatant piece of obscurantism. 'What is *really* driving the system?' they will demand. The implication is that if, in an experiment, a factor was varied and produced an effect, then surely that was *the* cause.

To assume that a given cell or a given condition is doing all the work may be good practice when setting up analytical experiments. However, any strong claim that one event is normally sufficient for the occurrence of another event will meet with frank incredulity from most ethologists.

Nevertheless the retort from the reductionists has been: 'If everything is as complicated as that no progress would have been made in any of the sciences that deal with systems'. Like many others (see Elman et al 1996), I have come to believe that the connectionist approach to neural and behavioural development will also form intellectual connections between the neural and behavioural scientists. It has to be said that many people who have adopted neural modelling are only interested in the outcome. They don't care if they have to invent *ad hoc* rules such as back propagation in order to get the thing to work. What they want is clever learning devices. However, if care is taken to develop models that are plausible at the neural level and do the right thing at the behavioural level, then we do have a ready made language for linking the levels.

References

Barkow JH, Cosmides L, Tooby J 1992 The adapted mind. Oxford University Press, New York

Bateson P 1987 Biological approaches to the study of behavioural development. Int J Behav Dev 10:1–22

Bateson P 1989 Additive models may mislead. Int J Behav Dev 12:407–411

Bateson P, Horn G 1994 Imprinting and recognition memory: a neural net model. Anim Behav 48:695–715

Bolhuis JJ, Johnson MH, Horn G 1985 Effects of early experience on the development of filial preferences in the domestic chick. Dev Psychobiol 18:299–308

Carey S, Spelke ES 1994 Domain-specific knowledge and conceptual change. In: Hirschfeld LA, Gelman SA (eds) Mapping the mind. Cambridge University Press, Cambridge, p 169–200

Caro TM, Bateson P 1986 Organisation and ontogeny of alternative tactics. Anim Behav 34:1483–1499

Chantrey DF 1974 Stimulus preexposure and discrimination learning by domestic chicks: effect of varying interstimulus time. J Comp Physiol Psychol 87:517–525

Elman JL, Bates EA, Johnson MH, Karmiloff-Smith A, Parisi D, Plunkett K 1996 Rethinking innateness. MIT Press, Cambridge, MA

Hinde RA 1969 Dichotomies in the study of development. In: Thoday JM, Parkes AS (eds) Genetic and environmental influences on behaviour. Oliver & Boyd, Edinburgh, p 3–14

Honey RC, Horn G, Bateson P 1993 Perceptual learning during filial imprinting: evidence from transfer of training studies. Q J Exp Psychol 46B:253–269

Honey RC, Horn G, Bateson P, Walpole M 1995 Functionally distinct memories for imprinting stimuli: behavioural and neural dissociations. Behav Neurosci 109:689–698

Horn G 1985 Memory, imprinting, and the brain. Clarendon Press, Oxford
Horn G 1991 Cerebral function and behaviour investigated through a study of filial imprinting. In: Bateson P (ed) The development and integration of behaviour. Cambridge University Press, Cambridge, p 121–148
Horn G, McCabe BJ 1984 Predispositions and preferences. Effects on imprinting of lesions to the chick brain. Anim Behav 32:288–292
Horn JM 1983 The Texas adoption project: adopted children and their intellectual resemblance to biological and adoptive parents. Child Dev 54:268–275
Johnson MH, Horn G 1986 Dissociation between recognition memory and associative learning by a restricted lesion to the chick forebrain. Neuropsychologia 24:329–340
Johnson MH, Horn G 1988 Development of filial preferences in dark-reared chicks. Anim Behav 36:675–683
Marr D 1982 Vision. Freeman, San Francisco
Plomin R, DeFries JC, Fulker DW 1988 Nature and nurture during infancy and early childhood. Cambridge University Press, Cambridge
Rowell CHF 1971 The variable coloration of the Acridoid grasshoppers. Adv Insect Physiol 8:145–198
Wu CS, Ganetzi B, Jan YN, Benzer S 1978 A *Drosophila* mutant with a temperature sensitive block in nerve-conduction. Proc Natl Acad Sci USA 75:4047–4051

DISCUSSION

Raff: You said that you are not ready for molecules. Is that because there's no way of getting at them? Don't you think it would help enormously if you could?

Bateson: It's an open question. Chip Quinn was talking this morning about his *Drosophila* work in which, working on mutants, he has been able to identify transmitters that make a big difference to the learning mechanisms. This might be true in our kind of work as well, but it hasn't yet been shown.

Perutz: If you cut sections of the brains of these chicks, do you see anatomical differences between the chicks that have learnt and those that have not?

Bateson: We see both an increase in connections and a loss of connections.

Rose: In the model we use, which is derived from the experiments I did with Patrick Bateson, you can see specific changes in synaptic conductivity that in different cases are both transient and lasting. I'm a little surprised that Pat says that molecules don't help, because we localized the IMHV site originally by looking at molecules.

Bateson: What I meant was that molecules don't help us understand the behaviour.

Rose: I think that's true. What they answer are the mechanistic question of how synapses are modified. This doesn't give you understanding of how that actually translates into different outputs.

Brenner: When a number of us became interested in behaviour and its connection with genetics, there were really two points of view. Seymour Benzer thought you could go straight from the genes to behaviour without worrying about the stuff in between. However, I think it is impossible to make a mapping of genes to behaviour which make any sense. The other approach is to say that there is something in between: namely, the genes are involved in building a nervous system that then behaves. There are now two almost separate problems: one is the problem of construction of the machine, which is essentially development, and the other is the problem of how the machine works, that is, physiology. These are two questions which can be reduced in different ways. Thus behaviour can be mapped onto the functions of the machine, and the genes map in a rather different way to the building of the machine. If you do not have this intermediate you cannot understand how genes might map on the behaviour, and it leads to absurd questions such as, 'Are there units of behaviour that correspond to genes?' The principle of construction is a necessary component of the explanation.

Gray: I want to put that into the context of what's going on at the moment in psychiatry, where there is at times a quite mindless link made between a gene and a very complex psychiatric condition, without any thought whatsoever as to what the gene might be doing to specify how the machine in the middle works.

Quinn: Sydney Brenner is partly right, but Seymour Benzer is not as wrong as you make him out to be. There are two defined areas in which Benzer's approach seems to have had a pay-off, one of which is learning and the other of which is circadian rhythms. He did eventually go into specification of eye morphology. On the other side, there has been a tendency to pick pieces of nematodes and call that neurobiology. For instance, the vulva has been touted as a model for the brain, but that is not necessarily true—and it's difficult to explain your work on the worm vulva to your kids!

Raff: I seem to remember Sydney Brenner saying much the same kind of thing about development—that there isn't a direct link between genes and development. But genetic analyses of development have transformed the way we think about development, making one optimistic that we'll be able to understand in detail how an egg develops into an adult organism. I suspect that the same thing will happen with our understanding of complex behaviour. It will fall into place bit by bit, and genetic analyses are likely to be an enormously powerful way of getting there. There have also been some important advances using simple organisms and biochemistry. There are some stereotypic behaviours that can be induced by a single peptide. Egg laying behaviour in *Aplysia*, for instance, can be induced by an injection of egg-laying hormone, which is normally released by bag cell neurons prior to egg laying. The behaviour induced is complex, and it is beginning to be understood in terms of the neurons involved and the actions of the hormone on them. This is comparable to the induction of metamorphosis in amphibians by an injection of thyroid hormone.

Garcia-Bellido: Martin, I think you are exaggerating now! Hormones have been known since the beginning of the century and they are not an 'explanation' of behaviour. cAMP causes changes to brain behaviour, but this doesn't mean that it explains it.

Raff: Don't you think it's a step forward to show that a particular peptide gives you the entire behaviour? You know the cells that make the peptide, as well as the cells that respond to it and, in some cases, you know the cellular response to the peptide.

Nurse: It has to be a step forward, of course, but the genes only describe the machine, as Sydney has said. How simple the connection is between a molecule and a behaviour depends on the system under study. It may be a rather simple connection or it may be immensely complex.

Wolpert: There is a big difference between development and the behaviour we're talking about here. In development, on the whole, local groups of cells do not interact with other groups of cells — everything is very local. In behaviour and in memory you've got processes spread over very large numbers of cells.

Gray: You have to distinguish between triggering a whole set of changes within a complex system, and that set of changes. If I yelled 'fire' and the fire alarm sounded, I dare say I could predict in advance a lot of behaviour. Would it be an explanation of that behaviour to say that I yelled 'fire'?

Raff: Exactly the same things were said about development. Incredibly complex processes are involved, and the actions of genes can be very indirect. Yet we now understand development in a way that no one would have predicted 10–12 years ago.

Maynard Smith: If I switch on the television set and find a beautiful picture, it doesn't tell me how the television works. It is helpful to know how to switch the bloody thing on, though.

Wolpert: This all brings us back to the question of what we mean by an explanation.

Noble: You hesitated at the end of your presentation to say that what you were referring to is antireductionist. I imagine you would be much happier to regard it as integrationist. An important distinction is that there are two different ways of looking at the alternatives to reductionism. One is to refer to antireductionism, and I think a good example of that was the philosophical point that I made after Horace Barlow's presentation (Barlow 1998, this volume). Reductionism and integration, though, are not incompatible. On the contrary, they are two facets of science that reinforce each other. What I presented yesterday in my work, was an example of integration as distinct from antireductionism (Noble 1998, this volume). I think that what you are doing is also best described as integration rather than antireductionism.

Barlow: Looking upwards is just as important, but perhaps less fashionable than looking down. I think there really is a difference, which is as follows. We are looking for explanations, and to my mind, an explanation has to be shorter than a description. You have to get some economy, and you get this by using information you already have. When you're looking downwards you can build your explanations on a sound foundation of physical and chemical knowledge. Building up on this enables you to give a briefer description of what's going on and how the lower level then explains the higher one. You don't get that benefit in the other direction: we don't have a firm background knowledge about social and other reactions at these higher levels. So there's an essential superiority in that sense in the reductionist approach: you are building on a platform of established fact and that enables you to give something which really is an explanation in the sense that it's a briefer account of the phenomena you're interested in than the description itself.

Noble: I totally agree: we have at the moment a very much firmer base at the physicochemical and molecular level. As things are at the moment, the arrows go much more easily in one direction than another. The point I was trying to make is that this is a cultural phenomenon: things could conceivably be very different in a few decades time.

What we identify as existing entities in the world is not simply an empirical matter. It is also a function of the way in which we see the world because our theories define and discriminate at least some of the things that exist. The part of the world that we have so far identified as existing (and of course we have to use all of the usual rules of science to ensure that these are not just figments of imagination) is itself a cultural thing; i.e. the part that we have chosen to study is a question of culture. It is easy to give examples of reductive entities: atoms, particles and the genetic code. It's also easy to give a lot of higher level ones: entropy, information, programs and social and economic interactions. We can even take the concept of love; some languages didn't use the expression 'I love you' until they imported it from western popular culture. Even the concept of a person doesn't exist as we know it in some languages, where individuals are defined in terms of society rather than the other way round. (If you want an example of the point I'm making about the things that exist depending on the way in which we see them, that's a very good one.) The danger is that in the reductive mode of thought we cease to see what still might be there to discover.

Morgan: Hubel may have been exaggerating when he said that one day we might be able to do without the Mind, but we could argue a solid case that the frontiers of mental explanation are retreating all the time. To take an example from perception, no less a person than Helmholtz, over 100 years ago, thought we should have to find what he called a 'psychic' cause for stereovision, because he couldn't envisage a way that the pictures from each eye could be put together optically or neurally.

Now, everyone thinks that stereovision is a relatively low level process carried out by neurons probably in V1. One could invoke many other examples. Perhaps there will always be some residual room for psychic explanations, but on the evidence we have it is likely that this room will shrink increasingly, because on the whole it represents a confession of ignorance.

Gray: The notion that the scope for mental explanation is shrinking only holds water if you have an implicit dualist understanding of what the 'mental' is. Of course, if you are a dualist, the mental will increasingly shrink, because dualism doesn't stand up. But that's not what 'mental' is within a scientific context— 'mental' is something to do with solving cognitive problems, and that can only be done by the brain. When you come up with an explanation of how the brain does it, you haven't shrunk the area of the mental, you have just understood the area better.

Morgan: So to clarify: you would say that stereo fusion is a mental process.

Gray: Yes, unless you're making that other distinction, between conscious and unconscious mentation. Let's not make the mistake of supposing that the mind and consciousness are the same thing: the great bulk of what we normally think of as mental operation is done unconsciously, and there is nothing in principal mysterious about how the brain does it. As regards conscious experience, we don't at present know how this fits into the world of science, and nobody has shrunk this area at all yet.

Williams: The basic argument here concerns what the components and the variables are, and the degree to which we can sort them out. I'll give you a very simple analogy. If we discuss the water molecule as a single molecule, then we say it is H_2O and we do not bother about any collective properties. The wave-mechanical approach I described earlier would describe the water molecule as far as anybody is concerned. That is a single particle description—it does not have variables associated with it such as temperature and pressure. Now if we describe an assembled molecular system, such as a block of ice, then what we may need to know about the block of ice is its temperature. However, it doesn't exist after a certain temperature—0 °C at atmospheric pressure. There are variables in the problem for many molecules which are different in what I call interactive systems from those where you deal with single molecules. So if you are dealing with single genes it could be like dealing with a water molecule, but if the system under discussion is a set of genes, then it is like a block of ice at various temperatures: it's got a cooperative entity associated with it which involves a different set of variables. You're talking about variables using the same units but which will be different in behaviour when you use different numbers of units, one or many water molecules, one or many genes.

Rose: Horace Barlow said earlier that explanations are short, descriptions are long, and what we're after is explanations. Denis Noble then offered us a whole series of things which are 'higher level' and which he maintains are phenomena that we can't talk about easily, such as love and personhood. I want to insist that if I say 'I am in love' or 'I am angry', then that is an explanation of my behaviour which is a good deal simpler and shorter than specifying the neuronal or hormonal state in my body at any time. Therefore it is crucially useful we keep this level of explanation.

Nagel: I am doubtful that the impulse towards reduction (which doesn't necessarily have to completely displace higher-level explanations, but which always seeks a deeper explanation of what we can explain on a more complex or larger scale, in terms of more basic components) is a cultural phase which might be reversed by a move towards the acceptance of certain higher-order explanations as fundamental. I actually think that the reductionist impulse conforms to a deep need for understanding — what we think of as really understanding something. So my expectation is that the things which we now can't reduce — and I would include consciousness in this — will lead us to search for new kinds of reduction. If consciousness can't be analysed entirely behaviourally, it may require ascribing some properties to matter that are capable of giving rise to it in sufficiently complex forms of organization.

Bateson: There is a perfectly orderly way of moving from an acceptance that another human has consciousness to investigating whether other animals might also be conscious. We can do so by looking at the behaviour and the physiological states that are associated with a particular condition such as damage to the body. We can then ask whether other animals show the same behavioural and physiological changes when they are put in a similar situation that would cause humans pain. This process is not as opaque as is being suggested here.

Garcia-Bellido: Why do we pass a judgement now? We are currently in the process of understanding. To say that the system doesn't work because at the moment we haven't described it is naïve. It is the attitude that counts, not just whether you have solved it completely today.

References

Barlow H 1998 The nested networks of brains and minds. In: The limits of reductionism in biology. Wiley, Chichester (Novartis Found Symp 213) p 142–159

Noble D 1998 Reduction and integration in understanding the heart. In: The limits of reductionism in biology. Wiley, Chichester (Novartis Found Symp 213) p 56–72

What is wrong with reductionist explanations of behaviour?

Steven Rose

Department of Biology, The Open University, Milton Keynes MK7 6AA, UK

Abstract. Methodological reductionism has served biology well, but its problems in the study of behaviour include turning open systems into closed ones, defining the units of analysis, and interpreting correlative and causal relationships between processes studied within different biological discourses, from molecular biology to psychology. The problems become more acute when methodological becomes philosophical reductionism, with its declared goal of collapsing 'higher level' explanations into 'lower level' ones. Quite apart from the vexed question of what constitutes a 'level', relevant behavioural phenomena may only be manifest at such higher levels. The reductionist programme assumes that parts have ontological and possibly historical (developmental, evolutionary) primacy over wholes, yet the nature of living systems is such that this cannot be the case. I will exemplify these problems in the context of the study of behaviour. But the worst problem arises when reductionism becomes an ideology, especially in the context of human behaviour, when it makes the claims to explain complex social phenomena (e.g. violence, alcoholism, the gender division of labour or sexual orientation) in terms of disordered molecular biology or genes. In doing so, ideological reductionism manifests a cascade of errors in method and logic: reification, arbitrary agglomeration, improper quantification, confusion of statistical artefact with biological reality, spurious localization and misplaced causality.

1998 The limits of reductionism in biology. Wiley, Chichester (Novartis Foundation Symposium 213) p 176–192

Reductionism is a portmanteau word, in which are packed multiple meanings, many of which have been discussed at this Symposium. I begin by distinguishing between reductionism as methodology, as philosophy and as ideology. Whilst I have problems with the first two, the burden of my criticism is focused on the slippery slope which leads both towards ideology (Rose 1997).

Reductionism as methodology

The living world is characterized by dynamic complexity. But we find it easier to understand phenomena, whether the behaviour of molecules or of animals, if we can isolate them from the rest of the world and alter potential variables singly. It is hard to make sense of what you observe if several features of a system are changing simultaneously. Reductionist methodology simplifies, and enables one to generate seemingly linear chains of cause and effect. It is no surprise that it has been both powerful and attractive over the last 300 years, providing unrivalled insights into the mechanics of the universe, because it often seems to work. Experiments are productive, findings replicable, predictions about the world are confirmed.

But living systems are not simple. They involve many interacting variables. Parameters are not fixed. Properties are non-linear. And the living world is highly non-uniform; the exception is nearly always the rule. So if one is not careful, the simplifying constraints that the methodology offers soon cease to be helpful supports to theory, and instead become straightjackets. What happens in the test-tube may be the same, the opposite of, or bear no relationship at all to what happens in the living cell, still less the living organism in its environment. It all depends.

Philosophical reductionism

The philosophical reductionist view is that because science is unitary, and because physics is seen as the most fundamental of the sciences, then an ultimate 'Theory of Everything' will be able to reduce chemistry to a special case of physics, biochemistry to chemistry, physiology to biochemistry, psychology to physiology, and ultimately sociology to psychology and hence to physics. But is such a reduction really even theoretically possible? Consider the relationship between physiology and biochemistry, exemplified by the study of muscle. Physiology studies muscle contraction, biochemistry the molecular processes that occur during this contraction. The biochemistry of this process is pretty well understood down to some of the minutest molecular details (see Holmes 1998, this volume). So why can't we just replace the physiologist's statement about muscle contraction with a statement about actin, myosin, etc.?

If the purpose of doing so is to claim that the biochemistry is causally responsible for the physiological event, this is a very different use of the word 'cause' from the way it is normally employed to describe a relationship in time between cause and effect, in which the proximal cause of the muscle twitch is provided by the physiological description of impulses travelling down a motor nerve to the muscle. The biochemical process does not *precede* the muscle contraction; it *describes* the muscle contraction. We are really making not a causal but an *identity* statement. But the physiological and biochemical descriptions serve quite different and complementary purposes. The biochemistry of the muscle twitch does not occur in the isolation of a test-tube. Muscles are anatomical structures, with specific locations within the organism. The relationships between molecular processes are organized in space and time in a manner which is not implicit in their chemistry. The *meaning* and *function* to the organism of the muscle twitch is apparent in the physiology and anatomy, but quite absent from the biochemistry. It cannot simply be eliminated.

Each 'level' of complexity of nature involves new interactions and relationships between the component parts which cannot be inferred simply by taking the system to pieces. Yet philosophical reductionism implies that even if higher-order properties are emergent they remain secondary to lower-order ones. The lower the order the greater the primacy. This assumption has dominated many of the presentations at this Symposium, from the taken-for-granted approach to muscle proteins (Holmes 1998, this volume) and Lewis Wolpert's view that 'this is what we feel comfortable with', to the more formal philosophical account by Nagel (1998, this volume). Indeed for him it seems as if only lower order explanations can be 'truly' scientific. Parts must come before wholes. Yet the nature of evolutionary and developmental processes in biology means that there is no such necessary primacy. Wholes, emerging, may in themselves constrain or demand the appearance of parts.

Reductionism as ideology

From its inception, science has been about both knowledge and power — above all, the power to control and dominate nature, including human nature. Thus the dramatic advances in knowledge of the past decades have been accompanied by ever more strident reductionist claims that the new genetics, molecular biology and neuroscience are about to first explain, and in due course to modify, the human condition (Rose 1995). It is here that reductionist methodology and philosophy tip over into ideology.

At its simplest, reductionist ideology argues a directly causal relationship between gene and behaviour. In a social and political environment conducive to such claims, and which has largely despaired of finding social solutions to social problems, apparently scientific assertions become magnified by press and politicians. Claims to explain phenomena as diverse as sexual orientation, mental distress or violence on city streets are scarcely minor concerns. We all want to know where to look to explain our personal successes and failures, our foibles and vices, to say nothing of the chronic crises of the world around us. If the causes of our pleasures and our pains, our virtues and our vices, lie predominantly within the biological realm, then it is to neurogenetics that we should look for explanation, and to pharmacology and molecular engineering that we should turn for solutions. This simplification is based on a faulty reductive sequence whose path the next sections trace and exemplify from the literature of studies on violence, sexual orientation and intelligence. It is not necessarily the case that any individual step in this sequence is inevitably in error, it is just that each is slippery and the dangers of tumbling very great. The issue at stake is the question of the appropriate level of organization of matter at which to seek causally effective determinants of the behaviour of individuals and societies.

Reification and agglomeration

Reification converts a dynamic process into a static phenomenon, a phenotype. Violence is the term used to describe dynamic processes of interactions between persons, or even between a person and their non-human environment. Reification transforms the process into a fixed thing—'aggression'—which can be abstracted from the interactive system in which it appears and studied in isolation. Yet if the activity can only be expressed in an interaction between individuals, to reify the process as if it were an isolable character attached to an individual is to lose its meaning.

Arbitrary agglomeration carries reification a step further, lumping together many different reified interactions as if they were all exemplars of the one character. Thus *aggression* becomes the term used to describe processes as disparate as a man abusing his lover or child, fights between football fans, strikers resisting police, racist attacks on ethnic minorities, and civil and national wars. Agglomeration proceeds by assuming each of these social processes is merely a reified manifestation of some unitary underlying property of the individuals, so that identical biological mechanisms are involved in, or even cause, each. This is well illustrated in the paper by Brunner et al (1993) describing a Dutch pedigree including eight men 'living in different parts of the country at different times' across three generations who showed an 'abnormal behavioural phenotype'. The types of behaviour included 'aggressive outbursts, arson, attempted rape and exhibitionism'. Can such widely differing types of behaviour, described so baldly as to isolate them from social context, appropriately be subsumed under the single heading of 'aggression'?

Much attention was devoted to the report that each of these 'violent' individuals also carries a mutation in the gene coding for the enzyme monoamine oxidase (MAOA), which, amongst other functions, is associated with neurotransmitter metabolism and is believed to be the site of action of a number of psychotropic drugs. Could this mutation then be the 'cause' of the reported violence? Brunner himself subsequently disclaimed the direct link and, indeed, dissociated himself from the public claims that his group had identified a 'gene for aggression' (Brunner 1996). Yet the paper is now widely cited in the research literature, and what Brunner described as 'abnormal' now becomes 'aggressive' behaviour. Thus a paper with this title, describing mice lacking the MAOA enzyme, appeared two years later (Cases et al 1995). The authors describe the mouse pups as showing 'trembling, difficulty in righting, and fearfulness . . . frantic running and falling over . . . (disturbed) sleep . . . propensity to bite the experimenter . . . hunched posture . . . ' Of all these features of disturbed development the authors chose only to highlight aggression in their paper's title, and to conclude their account by claiming that these results support 'the idea that the particularly aggressive behaviour of the few known human males lacking MAOA . . . is a more direct consequence of MAOA deficiency'.

As with each step in the reductionist cascade I am describing, the problem lies not in the fact that as researchers, within the methodology available to us, we need to classify or group together different types of observation as belonging in some way together. These are not inevitably illegitimate steps. Science often proceeds by alternately grouping together different phenomena as aspects of the same (lumping) and recognizing differences between them (splitting). Lumping is however inappropriate as applied to 'violence' as in these examples. Grouping arson and exhibitionism in the same category is not likely to make much sense to either a criminologist or a judge and jury in court.

To get round this difficulty, some researchers have recently re-labelled these cases, so that they no longer appear as examples of 'violence', but of a different category, of 'antisocial behaviour' (Rutter 1996). But such re-labelling only makes it worse. Just as agglomeration lumps disparate activities, so the identical act may be regarded as socially acceptable or unacceptable depending on the circumstances. Bombing a government building if you are a pilot and your nation is at war with those you are bombing is socially praiseworthy; on the other hand if you are a national of the society whose buildings you bomb you are guilty of the antisocial behaviour called terrorism. Contrast the medals given to US pilots during the Gulf War with the criminal charges against the bombers of the Federal office in Oklahoma City.

Improper quantification

Improper quantification argues that reified and agglomerated characters can be given numerical values. If a person is violent, or intelligent, one can ask how violent or how intelligent by comparison with other people. This assumption that any phenomenon can be measured and scored is reflective of the belief that to mathematicize something is in some way to capture and control it. The best-known example is IQ. 'Intelligent behaviour', essentially an interactive process between an individual and others, or with the social, living and inanimate world, becomes fixed as a unitary character (reification). Many different examples of such behaviour are then all taken to be manifestations of something called — as if finally to freeze dynamics into statics — 'crystallized intelligence', and given a special symbol, g, originally introduced by Spearman in the 1920s (agglomeration). Tests are then devised to measure this inferred hidden constant. The extraordinary belief is that all the multiple aspects of behaviour that go to comprise what we may recognize as intelligence — speed and accuracy of responding to new information, skill at deriving meanings from ambiguous social situations, capacity to innovate in novel environments, and many others as well — can all be reduced to a single number, so that the entire human population can be ranked, just as they might be if we were to line them all up by height.

To see this cavalier rejection of anything other than their own reduction of intelligence at its most arbitrary, one need go no further that the first chapter of Herrnstein & Murray's (1994) *The Bell Curve*, which, faced with voluminous critiques from many different perspectives of such reduction of intelligence to a single score, insists that intelligence is not to be confounded with talent, insight, creativity, or capacity to find or solve problems or resolve difficulties, any more than it has anything to do with musical, spatial, mathematical or kinaesthetic ability, sensitivity, charm or persuasiveness: it is reduced to the single number intelligence tests provide.

Statistics and the norm

Belief in statistical normality assumes that in any given population the distribution of such behavioural scores is Gaussian. Again the best known example is IQ, the tests for which successive generations of psychometricians refined and remoulded until it was made to fit (almost) the approved statistical shape. That is, tests which did not result in distributing the population according to the curve were rejected or modified, until they fitted the curve, a feat achieved between the wars in the various revisions of what became known as the Stanford-Binet, originally developed in the 1920s. Yet the assumption that the entire population can be normally distributed along a single dimension is to confuse a statistical manipulation for a biological phenomenon.

The power of this reified statistic should not be underestimated. It conveniently conflates two different concepts of 'normality'. The statistical sense of the term does not have a 'value' attached to it, it merely describes a particular shape of curve which has the property that 95% of its area is to be found within two standard deviations of the mean. But in common parlance the term normal is indeed normative. It describes not merely how things are, but how they ought to be; to lie more than two standard deviations from the mean in a Gaussian distribution is to be *abnormal*. When Herrnstein & Murray (1994) called their book *The Bell Curve* they played precisely into these multiple meanings of reified normality.

Spurious localization

Having reduced dynamic processes to reified numbers, they cease to be a property even of the individual, but instead become that of a part of the person. So the penchant for speaking of, for example, schizophrenic brains, genes — or even urine — rather than of brains, genes or urine derived from a person diagnosed as suffering from schizophrenia. Of course, everyone knows that this is a shorthand, but the resonance of 'gay brains' or 'selfish genes' does more than merely sell books for their scientific authors; it both reflects and endorses the modes of thought and explanation that constitute neurogenetic determinism, for it disarticulates the complex properties of individuals into isolated and localized lumps of biology.

Thus recent years have seen an unusually polemical debate, more reminiscent of the early days of 19th century phrenology than of modern research, amongst different neuroanatomists each claiming to have found 'the' brain seat of homosexuality. Two regions in particular have contended for the honour of conveying male same-sex preference: the corpus callosum and the hypothalamus. The bases for such claims have been subject to detailed and stringent empirical criticism by Fausto-Sterling (1992). My concern here is with the structure of the argument deployed by those seeking to locate homosexuality in a bit of the brain or an aberrant gene, for it shows all the features I have already described for violence and intelligence, and more besides. The expression of same-sex preference is scarcely a stable category either within an individual's lifetime or historically — indeed, that it might be used as a term to describe an individual, rather than part of a continuum of sexual activities and preferences available to all, seems to have been a relatively modern development. What the reductionist argument does is to remove the description of sexual activity or preference as part of a relationship between two individuals, reifies it and turns it into the phenotypic 'character' resulting from one or more abnormal brain structures. As always, it deprives the term of personal, social or historical meaning, as if to engage in same-sex erotic activity or even to express a same-sex preferred orientation meant the same in Plato's Greece, Victorian England or San Francisco in the 1960s.

Misplaced causation

It is at this point that neurogenetic determinism introduces its misplaced sense of causality. During sexual or aggressive encounters people show dramatic changes in, for instance the levels of circulating steroid hormones and adrenalin in their bloodstream and the release of neurotransmitters in their brains, all of which can be affected by drug treatments. People whose life history includes many such encounters are likely to show lasting differences in a variety of brain and body markers. But to describe such changes as if they were the *causes* of particular behaviours is to mistake correlation or even consequence for cause. When you have a cold, your nose runs. Yet despite the invariable correlation of the two, it would be a mistake to believe that the cold was caused by the nasal mucus; the chain of cause–effect runs in the reverse direction. Nor, despite the fact that

Prozac both inhibits serotonin re-uptake mechanisms and may diminish the likelihood of you committing suicide or murder, does this mean that the level of serotonin release in your brain is the cause of your desire to kill yourself — or someone else? After all, when one has toothache one can alleviate the pain by taking aspirin, but it does not follow that the cause of the toothache is too little aspirin in the brain. This issue has dogged interpretation of the biochemical and brain correlates of psychiatric disorders for decades, yet it still continues.

Dichotomous partitioning

If aggression, antisocial behaviour, or homosexuality are 'caused' by some 'abnormality' in brain structure or biochemistry or hormonal imbalance, what 'causes' these in their turn? They could of course be the consequences of some feature in the environment. More often, though, attention turns to those well-known first causes, the genes, and the apparatus of heritability studies is wheeled out. For even if there is difficulty in regarding such socially defined attributes as simple phenotypes, if they correlate with a 'real' measure such as the level of an enzyme or neurotransmitter, then the heritability of this can surely be determined. A good example of this mode of thinking is the claim that IQ test scores correlate with a more neurophysiological measure referred to as 'inspection time', whose heritability can then be assessed. The fact that a heritability measure is rarely applicable to the human situation, is widely misunderstood and (in most cases) meaningless has not prevented behaviour geneticists and psychometricians endeavouring to apply it, nor deprived it of its ideological resonance, as when it is reiterated that the heritability of intelligence — or rather of IQ test score — is as high as 80%. Political orientation, neuroticism and attitudes to military drill, royalty, censorship and divorce, amongst many others, are all supposed to show relatively high heritability. Indeed it becomes hard to find any human attribute or belief, even the most seemingly trivial (Rushton 1994), to which the heritability statistics fail to yield significance. Newly sophisticated statistical techniques, such as that known as Quantitative Trait Locus analysis (Plomin et al 1994), are employed purporting to show that even those conditions for which major genetic causation cannot be shown are in fact the result of the small additive effects of many genes. (The sad truth is that there is no distribution of a phenotype to which, given enough assumptions about partial penetrance and incomplete dominance, a genetic model cannot be fitted.) And whilst no-one claims that heritability equals destiny, nor that the figure provides information about any specific individual, rather than measures variance within a population, none the less, the whole tenor of the approach is to transfer the burden of explanation, and if appropriate of intervention, from the social or even personal level to that of pharmacological or genetic control.

Confounding metaphor with homology

If first causes are genetic, it then becomes appropriate to seek for equivalents of the human behaviour under consideration in the non-human animal world—that is, to find an animal model in which the behaviour can be more readily controlled, manipulated and quantified. Place an unfamiliar mouse into a cage occupied by a rat and the rat is likely eventually to kill the mouse. The time taken for the rat to perform this act is taken as a surrogate for the rat's aggression; some rats will kill quickly, others slowly or even not at all. The rat which kills in 30 seconds is on this scale twice as aggressive as the rat which takes a minute. Such a measure, dignified as *muricidal behaviour*, serves as a quantitative index for the study of aggression, ignoring the many other aspects of the rat–mouse interaction, for instance the dimensions, shape and degree of familiarity of the cage environment to the participants in the muricidal interaction, whether there are opportunities for retreat or escape, and the prior history of interactions between the pair. These are not merely speculative variables; many have been studied in detail by ethologists and shown to profoundly affect the nature of the relationships between the animals. But the reductive procedure goes further, for it then assumed that, just as time to kill becomes a surrogate for a measure of aggression, so this behaviour in the rat is transmogrified into an analogue of the aggression shown by drive-by gangs shooting up a district in Los Angeles. That is if one can find physiological or biochemical mechanisms—brain regions, neurotransmitters or genes, associated with the so-called 'aggression' in mouse-killing rats—then there should be equivalent or identical brain regions, neurotransmitters or genes involved in human 'aggression' (Johnson 1996). This type of evolutionary fantasy at best confounds a metaphor or analogue with a homologue. At worst it simply makes a bad pun on different meanings of the word 'aggression'. But it has become the vital, ultimate link in the chainmail armour of reductive ideology.

The consequences of reductionist fallacies

Methodological reductionism has proved a powerful and effective lever with which to move the world. We owe to it many of the most penetrating insights into mechanisms in every field of science, including biology. But especially in biology, complexity and dynamics, open rather than closed systems, are norms rather than exceptions, and the methodology of reductionism however powerful, has difficulties in dealing with complexity—indeed it may be positively misleading.

Reductionism as an ideology, insisting on trying to account for higher level phenomena in terms of lower level properties, hinders biologists from thinking adequately about the phenomena we wish to understand. But two consequences at least lie in the social and political domain rather than the scientific. Reductionist ideology serves to relocate social problems to the individual, thus 'blaming the victim' rather than exploring the societal roots and determinants of the phenomena that concern us. Violence in modern society is no longer to be explained in terms of inner city squalor, unemployment, extremes of wealth and poverty and the loss of the hope that by collective effort we might create a better society. Rather, it is a problem resulting from the presence of individual violent persons, themselves violent as a result of disorders in their biochemical or genetic constitution.

The second immediate social consequence of reductionist ideology is that attention and funding is diverted from the social to the molecular. If the streets of Russia are full of vodka-soaked drunks, and rates of alcoholism are catastrophically high amongst native Americans or Australian aborigines, the reductionist ideology demands funding research into the genetics and biochemistry of alcoholism. And it becomes more productive to study the roots of violent 'temperament' in babies and young children than to legislate to remove handguns from society. The point is that, as the whole of my argument up till now has stressed, for any phenomenon in the living world in general and the human social world in particular, one can offer multiple forms of explanation. But for any such phenomenon there are also *determining levels* of explanation — those which most clearly account for its specificity and point to potential sites of intervention.

So whilst in an ontologically unitary universe it is axiomatic that there is something different about the biochemical and physiological state of someone who is in the process of committing a murder from those states in the same person when he is in a prison cell, and probably between the murdering individual and someone who in similar circumstances does not murder, this difference cannot be relevant to answering questions about the causes and responses to social violence. Nor, therefore, can it represent the appropriate level at which to intervene if we wish to reduce the amount of violence on the streets. A programme devoted to the detection of which levels of serotonin might predispose a person to an increased statistical possibility of engaging in one of a number of activities, from suicide through depression to murder, followed by the mass screening of individual children to identify at-risk individuals, their drugging throughout life, and/or raising in environments designed to alter their serotonin levels — which is after all the action programme that would result from an attempt to define the genetic/biochemical as the right level for intervention — only has to be enunciated to demonstrate its fatuity. Good, effective science requires a better recognition of determining explanation and hence the determining level at which to intervene. Failing this it becomes a waste of human ingenuity and resource, a powerful ideological strategy of victim-blaming and a distraction from the real tasks that both science and society require.

References

Brunner HG 1996 MAOA deficiency and abnormal behaviour: perspectives on an association. In: Genetics of criminal and antisocial behaviour. Wiley, Chichester (Ciba Found Symp 194) p 155–167
Brunner HG, Nelen M, Breakfield XO, Ropers HH, van Oost BA 1993 Abnormal behavior associated with a point mutation in the structural gene for monoamine oxidase. Science 262:578–580
Cases O, Seif I, Grimsby J et al 1995 Aggressive behavior and altered amounts of brain serotonin in mice lacking MAOA. Science 269:1763–1768
Fausto-Stirling A 1992 Myths of gender: biological theories about women and men. Basic Books, New York
Herrnstein RJ, Murray C 1994 The bell curve. Free Press, New York
Holmes KC 1998 Muscle contraction. In: The limits of reductionism in biology. Wiley, Chichester (Novartis Found Symp 213) p 76–92
Johnson HC 1996 Violence and biology: a review of the literature. Fam Soc: J Contemp Human Serv 77:3–17
Nagel T 1998 Reductionism and antireductionism. In: The limits of reductionism in biology. Wiley, Chichester (Novartis Found Symp 213) p 3–14
Plomin R, Owen MJ, McGuffin P 1994 The genetic basis of complex human behaviors. Science 264:1733–1736
Rose SPR 1995 The rise of neurogenetic determinism. Nature 373:380–382
Rose SPR 1997 Lifelines: biology, freedom, determinism. Allen Lane, Penguin Press, Harmondsworth
Rushton J-P 1994 Race, evolution and behavior. Transaction Publishers, New Brunswick, NJ
Rutter 1996 Introduction: concepts of antisocial behaviour, of cause and of genetic influences. In: Genetics of criminal and antisocial behaviour. Wiley, Chichester (Ciba Found Symp 194) p 1–20

DISCUSSION

Nagel: I think that the issue between you and Dawkins is fundamental. You, in particular, want to present evolutionary theory in a tamer form. Dawkins wants to present it as a kind of reductionism that undermines the teleological or functional explanations that of course we have to continue to give — that the frog muscle twitched because he saw a snake and wanted to get away. It is a difficult question (and I'm inclined to be on Dawkins' side about this) as to how undermining the theory of natural selection is for ordinary explanations in terms of purpose. There are two kinds of relations that reduction can have to a higher level explanation. It can confirm it, as physics confirms chemistry by derivation of the laws of chemistry from physics, or it can displace it in effect by showing that the real story is such that in this case purposive explanations of muscle twitches are systematically misleading, even though we can't do without them for the purposes of predicting and so on. I still don't see why you have refuted this.

Rose: You use the word 'real' and this troubles me, because it implies that only one level or one type of explanation is the real one, and that the others are provisional in some way. It is precisely this philosophical reductionism that I'm disagreeing with. I don't think we're going to do away with physiology, animal behaviour or psychology, for instance, simply because we can describe these things in physical or chemical terms. They are appropriate and real explanations. As for my disagreement with Richard Dawkins, I would say that evolution by natural selection is one important part of understanding the way that the world is, but not all evolutionary change is selective. There are constraints on evolution which depend on contingency and structure and so on, and the levels of selection are not only at the level of the individual gene, they're at the level of the genome, the organism, the population and the species. We have to look at the multiple levels at which these operate. If you want to say that everything operates only at the level of single individual selfish genes, then you're making a statement which is not adequately scientific but which is an exceedingly powerful piece of ideology.

Morgan: Both Steven Rose and Pat Bateson have worked tirelessly to try to educate us about the enormous complexity of gene–environment interactions. Those complexities do exist, but there's a danger in over-emphasizing and almost mystifying them, ignoring the fact that often the issues are quite simple. The following is an experiment which illustrates my point. If I press on the left side of my eyeball, I see a black spot in the right side of my visual field: it is on the opposite side to the stimulus because my brain takes account of the optical reversal by the lens. Have I learnt that, or is it genetically programmed? This is the experiment by Schlotdtman, who took children who were congenitally blind because of cataract and who therefore had never had patterned vision. When they were about 10 years old he pressed a pencil on the side of their eye and found that they showed exactly this reversal. This is therefore innately genetically programmed. You go on and on about the complexities of gene–environment interactions to a point where people become desperate, but there are some simple questions and simple answers.

Gray: I've been troubled throughout this meeting by the fact that I have been agreeing with everything Steven Rose has been saying, but that has now changed. Let me first of all pick up on the word 'ideological'. Steven, it seems to me that you are being pulled by your own politics and ideology into making what is actually a good case into an extremely bad one by virtue of the examples that you used. You are using bad science to illustrate your point. Up until now the debate about what counts as reductionism and what counts as non-reductionism has been fought on a terrain where the science on both sides is good. That's the terrain it ought to be fought on.

For example, with the nature–nurture issue, you claimed that there is an arbitrary division between the efforts that statistical geneticists make to separate between genetic and environmental variance. Let me give you a concrete example from work that I was involved in with Jonathan Flint in Oxford and a group in Colorado (Flint et al 1995). We were able to use statistical methods to partition genetic and environmental variance and we then went on to do a study looking for the quantitative trait loci (QTLs) that determine the genetic variance. We found that there were pleiotropic genes which accounted for behaviour on a variety of different measures. When we added up the variance accounted for by the individual QTLs that we were able to identify, it made up exactly (give or take non-significant differences) the amount of variance that the first statistical genetical analysis had shown could be attributed to heritable variation. Thus you can do this type of calculation in a way that is good science. However, you gave examples of what may well be bad science. So you appear to be driven by your own understandable desire not to tolerate the superficial gene-for-this and -that approach (which all of us would deplore), and you are using that emotion to knock down something which actually is perfectly good science if done properly.

Rose: All the examples I chose were published in reputable journals, assuming you regard *Science* is a reputable journal. If they are bad science, one has to ask why reputable journals are publishing bad science. Why does a journal such as *Science* publish a study on violence and monoamine oxidase activity based on a pedigree of eight males (Brunner et al 1993), when neither Pat nor I could get away with publishing a paper based on a study of eight chicks? I really don't think that in this whole discussion about preferences for particular types of explanation we can separate entirely how we feel the world is organized from how we actually interpret our experiments, and that applies to me just as much to anyone else. I do, however, regard the whole statistical apparatus that goes with QTLs in looking at behavioural traits, as rather akin to the Ptolemaic epicycles before Copernicus appeared. That is, there's a great danger of confusing a statistical artefact with a biological reality. When I look at the statistical manipulations that go into a lot of behavioural genetics, I get very suspicious that I'm being phased by the statistics.

Gray: But exactly those same statistical methods have been applied to hypertension (Jacob et al 1991). These studies first of all established QTLs and subsequently at least one of them has moved onto the identification of genes involved.

Rose: Do they help us understand determining causes? I don't think they do.

Bateson: I want to reject the charge of obscurantism. Neither Steven nor I object to identifying the sources of variation in behaviour. However, we do object to people then using this evidence as a form of explanation by implying some straightforward link between the source and the behaviour. What I would like us to do, and this is where the science becomes really interesting, is to take the sources of variation, and explore precisely how they relate to the outcome. In this way we come up with much more dynamic models of how the development works and how we relate the sources of variation to the behaviour. At that point the science becomes truly 'integrationist'.

Quinn: I don't like applying genetic findings in animal systems, such as my work, to human behaviour. Nevertheless, some of the people working in this area are doing reasonable things although they haven't got hard data yet. E.O. Wilson originally proposed a gene for homosexuality and said what such a gene might do selectively: at the time I thought that this was offensive and stupid because he had no data. Now, however, people are engaged in a serious, not yet successful, search for such a gene. Whether you like this or not, if you find such a gene then that's truth: you can use it to consider aspects of human behaviour — certainly not all human behaviour — but if people find genes for complex human behaviours which I bet is less than 10 years away, then it's going to be difficult to run from the truth.

Wolpert: There is a strong association between possessing the Y chromosome and criminality. I don't know how anyone can say that genes don't determine behaviour: just look at male–female differences. What do you feel about the Y chromosome, Steven?

Rose: Human males are overwhelmingly responsible as individuals and as groups for the violence in society. We need to understand this in a variety of ways, about the nature of the organization of society, the nature of male/female relationships, and undoubtedly something about biology, the hormonal status of individuals. What I object to is the argument that there is a linear relationship between having a Y chromosome and violence. If you want to know the determining reasons why there is violence in society, to argue that it is caused by the Y chromosome is false, because there are an overwhelming number of males who are not violent in this sense. You might talk about predispositions, but what we have to understand is the complexity of the relation. This is what Pat Bateson keeps coming back to and I don't understand why people find it so problematic.

Gray: Steven, you are making arbitrary divisions as though the first bit — the organization of society and relationships between males and females — is not itself part of the biology.

Rose: Everything is part of everything.

Hess: It might be useful to remember that, since Greek science aimed to explore nature and describe it in terms of causes and effects in order to understand, predict and resynthesize. Classical science taught us to reduce experimentally and to design simple experiments to arrive at a description of complex phenomena in mathematical terms as linear laws of nature, and Leonardo da Vinci told us the logic of such procedures. The extension of analytical mathematics of science to the logarithmic scale yielded a broad extension of the time and space scale for exploration. Today, we are dealing with open and highly non-linear processes of the utmost complexity — perhaps never reducible — and, with the help of numerical computer techniques as well as analytical techniques, we are trying to resynthesize by computer modelling the mechanisms we have discovered by breakdown biochemistry. We hope to arrive at common laws describing the rules of molecular and cellular competition, selection, destruction and regeneration.

With respect to the notion that the stability of a given system is only given because it is a dynamic process and not a structure, I would like to say that this is not generally true, but is a rather relative statement. The study of properties of the hierarchical scale in time and space in biology shows clearly that each level has its own characteristic stability and instability regimes sometimes coexisting, and one stable level carries another one. The question of the properties of globally coupled molecular or cellular populations in dynamical systems is currently being investigated. Theoretical studies of globally coupled non-linear oscillators show that transitions to coherent and clustered regimes characterized by identical states of all elements can indeed occur. This approach indicates that understanding of the collective dynamics of network structures might well be reached, bridging the gap between the properties of individual and the group (P. Stange, D. Zanette, A. Mikhailov & B. Hess, unpublished results 1997).

Nurse: I want to shift the discussion to the philosophical argument that Steven was making. To underscore this issue, where do you stop for satisfactory explanations? For me this question is central to this symposium, and I still don't have it clear in my mind. There is this issue of what is explanation and description, and increasingly we are finding that each of us is using these terms differently. There is this pluralism which is being argued (explanations can be seen at different levels, they all have equal validity) and there is this other opinion that somehow it is better or more satisfactory the further we go down the levels of hierarchy. We really have to deal with this, because there is a tension there. Why is it more adequate to go further down? Can we really ever understand in that way, or is the pluralist approach better? To give an example, describing cellular reactions in terms of molecular interactions is a goal that many of us have. Given that there are many such interactions, and it may become more dull and tedious to go into all of those details, would it ever be satisfactory to represent what goes on in terms of the information flow and components and how they interact, or will that be unsatisfactory?

Noble: I think you have brought us back to one of the essential issues that this debate has been about, and that I welcome. But why is it that we should be concerned about such tensions? In almost every other form of intellectual debate that humans engage in, we take it for granted that there will be such tensions. I think that they are a sign of a healthy scientific community; not something to be ironed out.

Brenner: I think it is a question of who is asking the question. When a famous bank robber was asked, 'Why do you rob banks?', he said 'Because that's where the money is'. If his questioner had been his priest or moral advisor, his answer is irrelevant, but it is a perfectly reasonable answer for a social survey. As a scientist, I ask questions in a particular way for the explanations that I'm seeking.

Holmes: The use of levels in description is clearly very powerful and natural, but the moment when this tends to go wrong is in pathology or medicine. That's when you have to break levels, and one of the motivations which drives us back to a molecular description is in fact pathology.

Bray: I remember Francis Crick once saying that the reason that the strength of the molecular approach is that you can make a statement such as, 'The distance between one carbon atom and the next is 1.5 Å', and then defend the statement with all sorts of evidence from different kinds of experiment. That is why it pays if you can come down to the level of molecules, because other levels such as proteins, organelles and cells don't permit statements to be made with the same degree of rigour.

Williams: In this discussion a mistake is being made about even what physics is in reductionist terms. The basic idea would be to look at the variables in a system and ask about them. If you have a single individual or a single molecule, then that is a level of reductionism which you can go to for analysing certain ideas, but it does not and cannot include collective properties of an ensemble. And so if you're using reductionism and you come down to a level where the collective property is the important property, it is stupid to go down to single molecules, because you cannot describe properties such as temperature, for example, in terms of single molecules. So if you try to describe the brain in terms of single synapses it's probably quite impossible because you have to think what are the variables of the brain. They may be a collective property of synapses. When people say 'I am going to describe the brain and its behaviour in terms of DNA', what they're saying is 'Please let me exclude everything else for the moment and I will do my best to describe a mental activity'. Somebody else will come back and say 'Right, what you've done is fine, but you have deliberately excluded all these other variables and I would like to know whether they're interactive'. If they're interactive, then the problem is a collective problem and not one which you can reduce to a discussion of DNA.

Rose: I have three brief points. Firstly, I suspect that the search for certainty is also somehow embedded in the Y chromosome. We live in an uncertain world. My second point regards the relationship between levels. One of the important things we haven't discussed is the possibility that there may be many relationships between levels; it's not necessarily one set of events at one level which translates to another. My third point is that my plea for epistemological pluralism is not a plea for 'anything goes'. Certain sorts of pseudoscience, whether it's astrology, religion or behaviourism, I regard as not going at all.

References

Brunner HG, Nelen M, Breakfield XO, Ropers HH, van Oost BA 1993 Abnormal behavior associated with a point mutation in the structural gene for monoamine oxidase. Science 262:578–580

Flint J, Corley R, DeFries JC et al 1995 A simple genetic basis for a complex physiological trait in laboratory mice. Science 268:1432–1435

Jacob HJ, Lindpaintner K, Lincoln SE et al 1991 Genetic mapping of a gene causing hypertension in the stroke-prone spontaneously hypertensive rat. Cell 67:213–224

Levels of organization in ecological systems

Robert May

Office of Science and Technology, Albany House, 94–98 Petty France, London SW1H 9ST, UK

Abstract. Broadly speaking, ecology seeks to understand the structure and dynamics of individual populations of plants and animals, of communities of interacting populations, and of ecosystems. Ideally — the reductionist dream — it would be nice to build such an understanding of how individual populations respond to disturbance upon a fundamental understanding of the behaviour and physiology of the constituent individuals. This is vastly ambitious. In practice, most successful applications of population biology, for instance to the management of harvested systems or to the control of pests and pathogens, have treated the population itself as the basic variable (in equations whose parameters are assessed phenomenologically, even though they are in principal derivable from the more basic parameters pertaining to the behavioural ecology of individuals). By the same token, studies of the structure and function of communities and ecosystems are often and usefully approached phenomenologically as things in themselves (topology of food webs, 'plumbing diagrams' of energy flows, etc.) rather than being derived in a more fundamental way from detailed consideration of the interaction among the constituent populations. Furthermore, at every level of approach, it is often difficult to perform meaningful experiments which control variables and isolate single factors in a tidy way. Frequently, the spatial or temporal scale is such that observational data and/or past records provide the only way to estimate parameters. In the 1970s and early 1980s this inability of much of ecology to conform rigidly to simplistic schemes of 'how science is done' caused much angst. In this paper I offer a grandly opinionated overview of these issues.

1998 The limits of reductionism in biology. Wiley, Chichester (Novartis Foundation Symposium 213) p 193–202

By way of introduction, I would like to air a couple of my long-held 'cocktail party-ish' thoughts relating to reductionism. In so far as there are levels of organization, there are interesting differences between the 'hard' and 'soft' sciences. On the whole, the so-called hard sciences are the easier ones, and the so-called soft sciences are the harder, more difficult ones. In physics, there are rules which operate largely independent of time or place, but in biology one is dealing with systems which are in a sense reactive so that the whole picture becomes hugely more complicated. In biology, one cannot assume the laws are the same at all places and at all times because they are hugely context dependent and the contexts change over evolutionary time. Moving beyond biology to the social sciences, it gets even worse: there are all the problems of the life sciences plus the fact that the entities under study are aware of what is happening and are modifying things in a conscious way.

This is associated with some other hierarchies. Most physicists have a contempt for philosophy (witness Richard Feynman), whereas life scientists meet occasionally to discuss philosophical issues, as here, and the social sciences to some people seem to consist mainly of arguing about methodology and philosophy. On another level, mathematics is simply taken as part of the fabric of physics, but there are still many biologists who consider it to be irrelevant to their field.

I have a reductionist instinct, in that I feel that ultimately everything is derivable from first principles. In a sense, everything can be deconstructed down to elementary particles. It can, of course, often be very difficult to do this. But more importantly, I don't see the point much of the time: this process, even if it were manageable, would often be entirely irrelevant. One case in point is the second law of thermodynamics, the knowledge of which (according to C. P. Snow) is the touchstone of an educated person. Interestingly, this law didn't emerge from some piece of 'blue skies' research, pursuing knowledge for its own sake. Rather, it came from attempts to build more efficient steam engines. Scientists working towards this very practical goal stumbled upon this fundamental law at a phenomenological level. Later on, people began to understand how the law could be derived from the underlying physics of molecular motion. Even then, this derivation contained elements of semi-mysticism, namely Boltzmann's constant. My point here is that although it is fascinating for us to know that thermodynamics can be derived from an underlying level, by and large this is uninteresting for people who are actually practising the application of thermodynamics.

Levels of organization in population structure

The behaviour of populations is inherently derivable from the behaviour of the underlying individuals. For example, if you want to understand how to manage a fishery, in principle you would like to understand the population on the basis of the underlying behavioural ecology of the individual fish, and thus build up a model from the lower levels. In fact, this usually is not what is done.

All that is necessary for most practical purposes is to treat the population as if it were an entity, and have equations for it with phenomenological parameters which pertain to the density-dependent effects at the population level. For example, in forestry one can deal with the three-halves thinning law and build elaborate and effective management models that are all cast in terms of the level of the population, even though the behaviour of the population (and the 3/2 law itself) is ultimately derivable from the level of the biology of the growth of constituent plants.

On the other hand, there are times when one wants to go a bit deeper without going all the way down. There are some complicated questions in ecology where populations fluctuate for reasons associated with underlying non-linearities. Recently developed methods enable us to say, for example, that we think that underlying the complicated dynamics are two basic kinds of effect ('two-dimensional chaos'), without us understanding exactly what these effects are. This is a curious, intermediate level of description. One example of this involves the population dynamics of lynx and hare in Canada, which is one of the longest time-series in ecology, going back about 150 years. There is a pronounced cycle with a period of about 11 years. There are two levels of understanding: on the one hand, the phenomenology that will enable one to predict next year's population and, on the other, the desire to understand what actually drives the cycle. Recently, an analysis has been carried out by Stenseth (1995), using these new methods, to show that the hare cycles seem to involve two biological variables, whereas for the lynx cycles only one variable appears to be involved. This analysis is essentially independent of details, but it can be interpreted as suggesting that for the hare the interactions with predators and with food supplies are both important, but the lynx cycles are linked only to those of the prey. This is an intermediate level of understanding. I find it encouragingly in accord with earlier speculations, based on plausible assumptions about the basic biology of the food–hare–lynx system (May 1973).

In short, I can give you examples of useful understanding at all levels, from those building up to the dynamics of the population and elucidating how the phenomenological parameters in a population-level model actually relate to underlying behavioural variables, through to situations in which one has only phenomenology, bridged by an increasing number of instances where one has a sense of how many variables are underlying the phenomenology. But, at the end of the day, for most of the purposes for which this work is being done, the phenomenological models are adequate. The problems we have in their implementation have to do with economic and sociopolitical factors, and these themselves are more understandable than is often recognized (Clark 1976).

For example, with food webs there is a fascinating body of work that looks at their structure and aggregates the data from many different studies (Pimm 1982, Cohen et al 1990, Polis & Winemiller 1996). There is one interesting level of study that says there is some sort of rough, phenomenological rule that on average each species is interacting with (eating or being eaten by) only four or five other species. Why are food chains not longer than they are? The total number of links that go from the plants to the top predator (typically humans, because we sequester

25–50% of all terrestrial primary productivity) is typically only three or four. Ecological textbooks say that this is because the efficiency of energy transfer imposes such limits. But there are problems with this simple explanation: if it were true, poikilothermic food webs would be longer because the energy transfer efficiencies are much higher than for homeotherms, and they are not. In exploring these general rules, one is documenting the patterns purely at this level. One can alternatively delve deeper, asking how the rules derive from the underlying interactions among the web's constituent populations. But for many practical purposes — such as worrying about the consequences of simplifying food webs, or looking at the effects of climate change on them — although that underlying understanding may be interesting, it might not be important.

Spatial and temporal scale

The whole idea that one can understand one level in terms of the level below it can become difficult in some areas of science because the appropriate scale — either spatial or temporal — for controlled experiments is too large or long. This is a practical question that has caused a great deal of anguish in ecology. This problem is not unique to ecology; it is also encountered in cosmology. If you want to understand the history of the Universe, there are limits to the scope for manipulative experiments.

In the USA, the National Science Foundation (NSF) went through a rather irritating hiccup from the mid 1970s to the mid 1980s when a bunch of people (who didn't understand physics but thought that ecology ought to be more like what they thought physics was) decided that the NSF ecology programme really ought to be doing reductionist things, and reductionism got confused with doing manipulative experiments. The outcome of this was an interesting phase: if you look at the manipulative experiments in ecological studies published over that time, 75% of them were on a spatial scale of less than 10 metres and 95% of them were on a time scale of five years or less, the time scale of a PhD thesis (May 1994). It is not at all clear that the most important questions in understanding community structure and response to disturbance necessarily happen on a scale of less than a metre or over a timescale of three years or less.

Cooperative phenomena and emergent properties

Another thing that makes biology inherently more complicated than physics, is that so much of biology is inherently non-linear. Much of the reductionist success in physics can be put down to the existence of linear superposition principles; one can dissemble things and then meaningfully reconstruct the whole by adding them back together. When you have non-linear phenomena, processes similar to phase transitions in physics occur, where small changes in one variable can make for discontinuous changes in the whole system. Phenomena such as these are difficult to intuit from the study at the lower level. For example, a small change in temperature can cause the transition from solid to liquid.

Let me give two biological examples. In fisheries collapses, suppose we did have a model of fisheries dynamics that was built on the study of individuals. We would need more than only an understanding of the behaviour of individual fish; we would also need understanding of how such individual behaviour affected the dynamics of entire populations of interacting individuals. The notion that a clear understanding of individual fish behaviour will, of itself, illuminate the threshold phenomena of fisheries collapse is as ridiculous as the notion that understanding molecules in modelling ferromagnetism will by itself yield insight into that cooperative phenomenon. This is something that is not always understood clearly in biology. Such occasional lack of understanding may have something to do with the discomfort some biologists have with mathematical and theoretical approaches. Going from ecology to immunology, I note that the belief of some that the immune system will be understood fully by an ever more brilliant descriptive understanding of how individual viruses interact with individual T cells or other components of the immune system is an understandable delusion. But it is exactly the same as thinking that understanding the behaviour of individual fish will illuminate the behaviour of the entire fishery. Interactions at the level of whole populations commonly have properties that are demonstrably non-linear, and consequently can produce wildly complicated outcomes.

Conclusion

In conclusion, I am an antireductionist in the sense that reductionism is often confused with saying all we need is brilliantly clever manipulative experiments that describe what is going on. I believe it is important to understand not just *what* is happening, but also *how* it is happening and, ultimately, *why* it is happening. If we are going to get philosophical, I am probably closer to the 'anything goes' approach than anyone else in this symposium. There is no one road to success in this enterprise, no instruction kit to guarantee you'll get the answer if you follow the recipe book.

Sometimes real understanding may necessitate looking at lower levels, building from physiology to individual behaviour to population dynamics to community structure. More commonly, phenomenological understanding of the level in question — be it individual or population or ecosystem — will be sufficient for most purposes, and too simplistic a reductionist programme will be doomed by difficulties of spatial or temporal scales, or uniqueness of history.

References

Clark CW 1976 Mathematical bioeconomics. Wiley, New York
Cohen JE, Briand F, Newman CM 1990 Community food webs: data and theory (Biomathematics, vol 20). Springer Verlag, New York
May RM 1973 Time-delay versus stability in population models with two and three trophic levels. Ecology 54:315–325
May RM 1994 The effects of spatial scale on ecological questions and answers. In: Edwards PJ, May RM, Webb NR (eds) Large-scale ecology and conservation biology. Blackwell, Oxford, p 1–17
Pimm SL 1982 Food webs. Chapman and Hall, London
Polis GA, Winemiller KO 1996 Food webs: integration of patterns and dynamics. Chapman and Hall, New York
Stenseth NC 1995 Snowshoe hare populations: squeezed from below and above. Science 269:1061–1062

DISCUSSION

Bray: To what extent does the phenomenological description of ecological patterns you described depend on large numbers of organisms? In epidemiology and biochemistry there are situations in which the system is reduced to small numbers of elements, in which case the descriptions that work perfectly well for large numbers no longer apply. Does that happen in ecology as well, and in that case are you not forced to use a more reductionist approach?

May: If you want to demonstrate that chaotic dynamics — simple rules that result in hugely complicated outcomes — are actually to be found exemplified by biological populations, there are several approaches. One is to bring populations into a laboratory and construct beautiful experiments where single populations are maintained, with all other variables kept constant. Then you could, for example, change the temperature, thus changing metabolic rate, and see chaotic dynamics. But whether that has anything much to do with understanding the dynamics of that particular population in the real world is an open question. What you have done in the laboratory is to abstract a piece of what is a real system from so much of the encumbering reality of the world that it is not clear to me that this approach is anything other than using a living computer to illustrate a mathematical possibility.

Bray: However, in epidemiological situations, populations do become small and infections of small numbers of organisms then become very influential.

May: One way of looking at this is to ask the question: for a given infectious agent, how small can a population be and still maintain the infection endemically? As you get to that population level, then of course individuals will matter if the level is small enough. On the other hand there are many interesting things you can say that rely on laws of large numbers and deterministic models. There isn't a blanket answer to what is appropriate.

Brenner: This emphasizes what I tried to say yesterday, about the machine language of the system that you're trying to simulate or understand (Brenner 1998, this volume). It is not necessary for the machine language to be embedded in the material of the level below. That is, you can have a machine language of the system that is effectively at the same level as the system itself. We need to define that in order to begin this task of understanding. As I mentioned yesterday, it would be lunacy for anybody interested in development to say 'I have really got to put this in quantum mechanical terms or it is useless'.

Quinn: Your description of population biology equations that work despite impoverished or insufficient or heuristic information sounds a bit like an applied science, such as modern chemistry or engineering. This is different from up-and-down science.

May: I think science is just about understanding what's going on, and I don't draw these rigid distinctions between pure and applied science. There are admittedly different levels of satisfaction in understanding a system. Understanding what's going on in, for example, a fishery, happens to be classified as 'applied'. I would argue that fisheries people have managed to obtain a fair understanding of what's going on the Grand Banks fisheries without having had to understand it in terms of larval recruitment.

Quinn: There's a bit more of a problem in the immune system in that there have been mathematical models of immune system function since the year dot and they have predicted correct phenomena. But until people knew most or all of the components to the immune system at some level they weren't dealing with the full deck. The models were delusory. Models without the information are in that case not solutions.

May: Models which are not based on minimal relevant sets of facts are clearly not very useful. I wasn't arguing for models as such: what I was saying was that the notion that one can understand the behaviour of certain kinds of system simply by understanding the behaviour of the constituent elements as such, without asking explicitly how the system behaves when large numbers of those elements interact, is foolish. But it would be even sillier to try to understand how large numbers of viruses interact with immune system cells without basing it on the individual interactions. An increasingly correct understanding of individual interactions is necessary as a basis for understanding how populations interact.

Raff: Immunology has been very successful in solving some of the major mysteries.

May: There's no question about that. If you ask me about the underlying facts on which you could then look at properties of aggregates, if you ask how important is that versus looking at the level of interacting populations, then it is 99:1. One can't begin to look at the population without looking at the individual interactions. But the recognition that this last little bit—populations of viruses interacting with populations of immune system cells—is important is denied by many of your colleagues.

Wolpert: What are your parameters for applying individual behaviour to a population?

May: Let me give you a concrete example. In the 1980s, the International Whaling Commission (IWC) used whale population models which were essentially of the form (here simplified somewhat):

$$N_{t+1} = (1 - \mu)N_t + R(N_{t-T}).$$

Here N_t is the population of (sexually mature) adult whales in year t. The corresponding population one year later, N_{t+1}, consists of the surviving fraction, $(1-\mu)N_t$, plus those newly recruited into the adult population from births T years ago, $R(N_{t-T})$; T is the time taken to attain sexual maturity. The models, incidentally, assume the sex ratio is unity, and that maturation times and mortality rates are the same for females and for males. The IWC models conventionally assume the recruitment term, $R(N)$, has the general form:

$$R(N) = \tfrac{1}{2}(1 - \mu)^T N[P + Q\{1 - (N/K)^z\}].$$

Here K is the pristine, unharvested equilibrium density of the whale population; P is the per capita fecundity of females at this pristine equilibrium point where $N = K$; Q is the maximum increase in per capita fecundity of which the whales are capable as population densities fall to low levels; and z measures the severity with which this density-dependent response is manifested. The factor $\frac{1}{2}$ expresses the fact that only half the population are producing offspring (remember, the sex ratio is unity).

In equations like this, many of the parameters have a direct biological significance, in terms of the average life history properties of individual whales: this is true of the death probability, μ; the average age of attaining sexual maturity, T; the average per capita fecundity at equilibrium, P; and the intrinsic maximum average fecundity, attained in the absence of density-dependent constraints, $P+Q$. The 'carrying capacity', K, also has a direct biological interpretation, although in practice the causative mechanisms may be obscure, and the actual magnitude of K may be hard to determine. In contrast with these life history parameters, with their direct biological meanings, the parameter z is the phenomenological fudge factor. Indeed, the assumption that the shape of the density dependent effects can be represented as $1-(N/K)^z$ is equally phenomenological; it is just a plausible function, parameterized by z, which can be forced to fit such data as are available.

Obviously, in the ideal world, one would like to — and indeed one could — derive the functional form of the density dependence from an underlying understanding of the biological factors which gave rise to it. But, at least for whale population dynamics, this is a counsel of perfection. We can, moreover, get a very good understanding both of the effect of harvesting upon the dynamics of whale populations (not only in direct changes, but also the indirect effects such as the changed ability to respond to environmental fluctuations), and more generally of the bioeconomics of the whaling industry, from such population-level equations. The reductionist would like to be able to estimate z in terms of a description of the behavioural ecology of individual whales, but many important practical problems as well as a fundamental understanding of the bioeconomics of whale harvesting can be gathered from these phenomenological models at the level of the population.

Kerszberg: A short remark about the notion of phase transition. This is usually seen as a source of trouble, because it leads to a discontinuity in the behaviour of the system. What makes possible the description of fish populations in terms of a very few variables and almost nothing relating to the fish itself is precisely an underlying phase transition. Let me give you a simple illustration in terms of a phase transition everyone here knows: liquid to solid. In the case of a liquid you have to know in principle the position of each molecule in the liquid, or you can settle for a statistical description of the positions. When you have a phase transition, this phenomenon is difficult to describe itself, but *beyond* the liquid–solid transition you have a lattice situation, a crystal, where all you need to know are the positions of the atoms in the basic motif which you just repeat forever. The change in description and its simplification is made possible because of the appearance of collective variables (what physicists call order parameters) and these order parameters can be used in describing the crystal in a very compact way. So phase transitions are what make collective descriptions possible at all, they are not just a nuisance.

Quinn: At some level I either disagree or am confused about mystical properties of the ensemble being different from the properties of individuals. It is true that formal thermodynamics is a discipline of limited reductionism, in that there is temperature and entropy and so forth. But one of the triumphs of turn of the century physics was the rederivation of those properties in terms of atomic physics, statistical mechanics and a bit of mathematics. So there are reasonable underpinnings of that relating it to individual atomic properties, and it seems like there are not mystical properties of whale populations except for this density dependence, which is derived.

May: I never suggested they are mystical properties, nor did I ever suggest that phase transitions, thresholds or other non-linear things that can only be understood when you look at whole populations of individuals are in the least mystical. All I am suggesting is that there is a certain kind of heads-down, narrow-focus mind set that says all you need do is focus on the interactions of individuals, and I think that approach is just too dim to appreciate some of the most important phenomena which connect one 'level' to the next.

Reference

Brenner S 1998 Biological computation. In: The limits of reductionism in biology. Wiley, Chichester (Novartis Found Symp 213) p 106–116

The units of selection

John Maynard Smith

School of Biological Sciences, Biology Building, University of Sussex, Falmer, Brighton BN1 9QG, UK

Abstract. Darwin's idea of evolution by natural selection is almost universally accepted by biologists, but debate continues about the units of selection. The history of this debate starts with Wynne-Edwards' arguments for group selection, and Hamilton's explanation of social behaviour in terms of the inclusive fitness of individuals. Hamilton's approach differs from the gene-centred approach pioneered by Williams and Dawkins, although both the problem and its solution are essentially the same. The choice of approach depends on conceptual and mathematical simplicity, and on one's attitude to the causal efficacy of genes. The problem of selection on units above the species level is discussed. Today, we are in the main concerned with cases in which selection acts simultaneously at two levels. This is true of current research on intragenomic conflict and of the suggestion by Maynard Smith and Szathmáry that in the major transitions in evolution, entities that were capable of independent replication before the transition can only replicate as part of a larger whole after it.

1998 The limits of reductionism in biology. Wiley, Chichester (Novartis Foundation Symposium 213) p 203–217

There is an almost universal acceptance of Darwin's theory of evolution by natural selection among biologists, but what are the units upon which selection acts? I will discuss this question by giving a brief history of the debates that have occurred during the last 50 years. This history will not be based on serious research, but on my own recollections. In particular, the account would probably be rather different had I been working during that period in the United States rather than Britain.

Prior to 1960, population geneticists ascribed finesses to genotypes (sometimes they referred to the 'mean fitness of the population', but this had nothing to do with the likely survival of the population, and was only a normalizing factor in equations). This amounts to treating the individual organism as the unit of selection. In contrast, among biologists as a whole the habit of explaining some trait in terms of the good of the species was not uncommon, although it was not always appreciated that such an explanation tacitly assumes that selection acts between species rather than between individuals. Matters were brought to a head by the publication in 1962 of Wynne-Edwards' *Animal dispersion in relation to social behaviour* (Wynne-Edwards 1962). The theme was that many patterns of behaviour, including territorial and lecking behaviour, had evolved because they enabled animals to regulate their population density before starvation intervened to do it for them. Wynne-Edwards' great virtue was that he recognized, and stated explicitly, that the mechanism he was proposing required selection between groups. The book was also beautifully written, full of fascinating natural history, and concerned with a topic—population regulation—which, reasonably enough, was at that time seen as central to ecology. For these reasons, it was an immediate success.

Not surprisingly, population geneticists were less enthusiastic, as were some naturalists, notably David Lack. In 1964, stimulated by Lack and Nicko Tinbergen, I wrote a paper entitled 'Group selection and kin selection' (Maynard Smith 1964). My main aim was to show that the process suggested by Wynne-Edwards required that a species be subdivided into a number of small populations with little genetic interchange. The process was not in principle impossible, but the necessary population structure did not exist in most of the species that Wynne-Edwards had discussed. While writing the paper, I became aware of the work of W. D. Hamilton (Hamilton 1963, 1964a,b). This seemed to me to be important. I did not want his essentially correct idea to be buried along with what I thought to be the erroneous idea of group selection. I coined the term 'kin selection' (my only contribution) to distinguish it from 'group selection': the essential distinction being that group selection, as proposed by Wynne-Edwards, requires the differential survival and reproduction of groups, whereas kin selection requires only that individuals interact with their relatives. I think, however, that the general rejection of Wynne-Edwards' proposal owed more to the arguments of Lack and others showing that the behaviours he had described could readily be explained by individual selection.

Although little noticed at the time, Hamilton's ideas have had far greater long-term impact. His central insight, that if organisms interact with their relatives, then genes causing altruistic or cooperative behaviour are more likely to increase in frequency, has been the basis of all subsequent attempts to understand the evolution of social behaviour. In the present context of 'units of selection', the important point is that Hamilton attempted to analyse the effects of relatedness while continuing to treat fitness as a property of an individual (or class of individuals with a specific trait), and not either of a family or of a gene.

I suspect that many more people have been influenced by Hamilton's 1964 papers than have actually read them, so I want to paraphrase what is in them (Hamilton 1964a,b). He defines two measures of the fitness of an individual. The first, 'neighbour-modulated fitness', is the expected number of direct offspring produced by an individual. It is this fitness that determines the fate of the genes carried by that individual. However, if an individual happens to carry genes causing altruistic behaviour, and if it interacts with relatives, then those relatives are more likely also to carry genes for altruism, and so are more likely to help it to produce offspring. Hence, in calculating fitness, we must allow for the help received from relatives — hence the term 'neighbour-modulated'. As Hamilton saw, it is horridly difficult to make this calculation. Instead, it is easier to calculate what he called 'inclusive fitness' — which in effect is a measure of the direct offspring produced by an individual, not allowing for the help received from neighbours, but adding any extra offspring produced by relatives of the individual because of the help it provides, appropriately weighted by relatedness. The justification for using inclusive fitness is, as Hamilton showed, that its use, subject to certain assumptions, leads to the same conclusions as the use of neighbour-modulated fitness.

I find it interesting that Hamilton went to such lengths to retain the notion of individual fitness, while allowing for the effects of relatedness. I suspect that he made this effort because he wanted to maintain the structure of population genetics. Today, however, most people would solve problems involving relatedness either by use of 'Hamilton's inequality', $b/c > 1/r$, or by a more explicitly gene-centred approach: I will return to this point below. For the moment, it is sufficient that Hamilton showed that a strictly individual-selection model was able to account for the evolution of social behaviour, even to the extent of sterile castes in social insects.

My own interest was stimulated by a different problem in animal behaviour. As a student I had read the work of Lorenz on the ritualized nature of some animal contests. Although I did not doubt his observations, I could not, as a student of J. B. S. Haldane's, accept the 'good of the species' terms in which they were explained. I was stimulated to revisit the problem by the suggestion by George Price (in an unpublished manuscript that I had seen as a referee) that ritualized behaviour in fights was to be explained by the possibility of retaliation. In an attempt to formalize his ideas, I was led to evolutionary game theory, and the notion of an evolutionarily stable strategy (ESS). Eventually, we published a joint paper (Maynard Smith & Price 1973) setting out our ideas. As in Hamilton's case, we proposed a model of individual selection, with the aim of showing that traits that had previously been explained by group selection could be explained without it; although in our case we made the simplifying assumption of asexual reproduction. Since that time a good deal of effort has gone into showing that, with few exceptions, sexual populations also will converge on an ESS.

We were not, in fact, the first people to use the idea of evolutionary stability. R. A. Fisher's theory of the 1:1 sex ratio is essentially that this ratio is the only evolutionarily stable one, although he did not (in this context) refer to the theory of games. Hamilton, however, in his 1967 paper on extraordinary sex ratios (Hamilton 1967), did explicitly seek an 'unbeatable strategy' when considering the sex ratio with 'local mate competition': I think that his idea was essentially the same as mine of an ESS. My only real claim to the idea is that I formalized it, and explored its range of application. Although originally developed to analyse animal contests (as the term 'strategy' implies) it has proved applicable in contexts as diverse as plant growth, and even the evolution of retrotransposons (Nee & Maynard Smith 1990), and is now the standard model for studying animal communication.

By 1975, individual selection had become the orthodoxy, and a lapse into group selection was liable to be met with hatred, ridicule and contempt. In science, however, it seems that the establishment of an orthodoxy is the signal for attempts to undermine it. These have come from two directions — attempts to re-establish the role of selection at levels above the individual, and attempts to replace individual by gene selection.

The first was heralded by D. S. Wilson's 1975 paper, 'A theory of group selection' (Wilson 1975). He considered a large, random-mating population which, each generation, breaks up into a number of 'trait groups', within which selection acts on individuals. The population consists of two types of individual, 'donors' and 'recipients'. He found that, if groups were formed randomly, then donors would increase in frequency only if the effects of their own actions increased their individual fitness: the influence of their actions on others is irrelevant. But if genetically similar individuals (which in practice would typically be relatives) assorted together into groups, then altruistic donors could increase in frequency. For some reason that still escapes me, he decided that the effect he had discovered was not caused by kin selection, and decided to call it group selection instead. Although his interpretation caused nothing but semantic confusion, I think the trait group model is of some value. If assortment into groups is random, and if (as Wilson assumed) costs and benefits combine additively to determine fitness, then it is easy to show that only an individual's effects on its own fitness matters, and interactions are irrelevant. The evolution of cooperation requires *either* that interacting individuals are relatives (or genetically similar for some other reason), *or* that fitness interactions are non-additive.

I think, therefore, that the real lesson of Wilson's model is that the evolution of cooperation requires either relatedness, or synergistic fitness interactions, or both. I will return later to the question of whether there are indeed levels above that of the individual that must be considered.

A more fundamental blow to the individual selection orthodoxy was struck by two books: G. C. Williams' *Adaptation and natural selection* (Williams 1966), and Richard Dawkins' *The selfish gene* (Dawkins 1976). There is some irony in this, since both books were intended primarily as a refutation of group selection ideas. But they do so from a new point of view — a gene-centred view. In fact, Hamilton had used this approach in his 1963 paper, but, in the fuller (and, apart from publication, earlier) 1964 treatment he ascribed finesses to individuals, not to genes (Hamilton 1963, 1964a,b). In using the phrase 'selfish gene', Dawkins was not, in the main, concerned with the possibility that particular genetic elements might multiply more rapidly than typical elements of the genome. This possibility was later raised, under the term 'selfish DNA', by Doolittle & Sapienza (1980) and Orgel & Crick (1980): I will return to it later. But Williams and Dawkins were considering well-behaved genes. Their point was that genes replicate (more precisely, the information they carry is replicated), whereas organisms do not. The problems they were interested in could be analysed using classical individual finesses (neighbour-modulated fitness), or by inclusive fitness, or by ascribing finesses to genes. Since in each case the assumptions about what was going on would be the same, the three methods should, if correctly applied in particular cases, give the same answer.

Why, then, does it matter? One answer is that some methods are simply harder to use than others. I have already mentioned the difficulty of calculating neighbour-modulated finesses. There are cases in which the easiest way to calculate is simply to ask 'how many copies of this gene will there be in the next generation?' I want to give a trivial example of this, because it raises some interesting points. This concerns sickle cell anaemia. Sober & Lewontin (1982) argued, as a reason for rejecting the gene-centred approach, that the polymorphism for the relevant gene could not appropriately be analysed in that way. I replied (Maynard Smith 1987) by pointing out that a gene-centred method was not only possible, but algebraically simpler: it leads directly to a first-order equation for the equilibrium frequency, instead of a rather messy cubic. Of course, cubics can be solved, and the answer is the same whichever method one uses. But I want at least to suggest that, if a method leads to simpler mathematics, that is a reason for preferring it. Of course, I am not claiming that reduction always leads to mathematical simplification: the opposite may be the case. For example, the justification for using evolutionary game theory to analyse the evolution of phenotypes whose fitnesses are frequency-dependent is that the corresponding genetic analysis is usually impossibly difficult. The gain achieved by modelling at the higher level arises because complications (in this case, sexual reproduction) can be omitted.

Perhaps a more important question is whether the gene-centred method is causally appropriate. Sober (1987), replying to my 1987 paper (Maynard Smith 1987), argued that algebraic simplicity is irrelevant, and that the gene-centred method must be rejected because the sickle-cell gene cannot be thought of as 'causing' the fitness of its carrier. His reason was that causes must be context-independent: Since the S gene increases organism fitness if its partner is A, but decreases fitness if it is S, it cannot be causal. This seems to me plain wrong. It would also rule out the individual selection model, because the fitness of an A/S heterozygote is context-dependent: it depends on whether there is malaria about.

In biology, all causes are context-dependent. When thinking about evolution, I often find it easiest to think, 'If I were a gene in that situation, I would do so-and-so'. That is, I think of genes as causal agents. Of course, such thoughts have later to be formulated mathematically, but gene-centred thinking is useful at the hypothesis-forming stage, as well as, sometimes, being mathematically convenient.

The present stage of the levels of selection debate arises from the discovery of 'selfish DNA'. The possibility that genes which cheat may have important evolutionary consequences has been recognized for a long time (for example, Hamilton's [1967] treatment of meiotic drive genes on the sex chromosomes). What is new is the realization that selfish genetic elements are widespread. The result has been the notion of intragenomic conflict. For example, the evolution of P factors in *Drosophila* (reviewed in Engels 1989): these are chromosome elements of 3000 bp, coding for two proteins, and able to multiply independently of chromosome replication. To understand why they do not increase indefinitely, until they have caused the extinction of the host species, one has to consider selection at two levels, within the individual fly, and between flies. The organism now plays the role of the group, and the P element the role of the individual.

This problem of selection acting simultaneously at different levels turns out to be central for the large-scale evolution of increasing complexity (Maynard Smith 1988, Maynard Smith & Szathmáry 1995). Although there is no reason, theoretical or empirical, to think that evolution necessarily leads to greater complexity, only a pedant would deny that dragon-flies and oak trees are more complex than any organism that existed 3000 Ma ago. This increase has depended on a rather small number of major changes in the way in which information is stored, expressed and transmitted between generations. Most of these transitions have a common feature: entities that were capable of independent replication before the transition can only replicate as part of a larger unit after it. There was therefore a conflict between selection at different levels. The point is best made clear by examples:

(1) The origin of 'compartments', or protocells. There would be both within-cell selection for molecules that replicated rapidly, and between-cell selection for molecules that cooperate to ensure rapid cell growth.
(2) The origin of chromosomes. Molecules initially capable of independent replication were linked together so that they could only replicate synchronously. Shorter molecules replicate faster, but linkage encourages the evolution of cooperation.
(3) The origin of eukaryotic cells involved symbiosis between a host cell and an endosymbiont, the mitochondrion (and, in algae, the chloroplast), which previously had been capable of independent replication.

(4) The origin of sex. Sexual populations gain in ability to evolve and in reduced mutational load, but individual cells could gain a short-term advantage by reverting to the ancestral condition of asexual reproduction.
(5) The origins of multicellular animals, plants and fungi. Most cells in such organisms have lost the ancestral ability to replicate independently.
(6) The origins of animal and human societies. In social insects, and in some other animals, most individuals have lost the ability to reproduce.

Two points can be made. First, even if we can point to selective advantages of the higher-level organism, its origin requires an explanation in terms of selection at the lower level. It is no good persuading ourselves that sexual populations have an advantage over asexual ones if it would always pay individual cells to multiply asexually. Second, an explanation of such transitions turns out to depend on two processes, the genetic relatedness of cooperating individuals (kin selection), and the more-than-additive advantages of cooperation (synergistic selection), which were discussed earlier.

Are there levels above that of the organism at which selection can be effective? The obvious candidate, in sexual organisms, is the species. Species are, in the main, reproductively isolated from one another, they go extinct, and they split into two. When they split, the daughter species resemble the parent. Thus they have the properties needed for evolution by natural selection. There is some evidence that species selection can be effective. For example, the taxonomic distribution of parthenogens suggests that they are selected against, in the long run, at the population level (Maynard Smith 1978). A study of mass extinctions suggests that marine species with planktonic larvae are more likely to survive than those with direct development. Two reservations are necessary. First, if there is widespread variation for a trait within a species, selection between individuals is likely to override selection between species: few species contain both sexual and obligately parthenogenetic females, and within-species variation for planktonic as opposed to direct development is likewise unusual. Second, it is pointless to invoke higher-level selection for traits that are expressed and selected in individuals. It is sensible to invoke colony-level selection for the traits that enable termites to build mounds, or species-level selection for traits that enable species to evolve rapidly, but the ability of *Drosophila* to fly does not call for selection above the individual level.

Are there relevant levels above that of the species? The existence of symbiotic unions suggests that there may be. However, the origin of such unions needs to be explained in terms of selection acting on individuals of each partner, even if, today, the union is so intimate that neither can survive without the other. I can see no case for regarding ecosystems as units of selection. I also think that the analogy drawn, in the Gaia hypothesis, between the biosphere and an organism is seriously misleading. It is important that the properties of the oceans and the atmosphere depend on the actions of living organisms, and that, in some cases, these properties are regulated by negative feedback. But the analogy with an organism implies the presence of specific receptor and effector systems which exist precisely because they ensure the survival of the biosphere. Since there is only one biosphere, it cannot evolve by natural selection, and hence the attribution to it of adaptive properties is unjustified.

There have been many controversies concerning the units of selection. Some have concerned different ways of modelling particular processes. A clear example is the debate over the effects of relatedness on the evolution of altruistic behaviour. Is this best modelled by a gene-centred approach, or by calculating inclusive fitness, or by classical fitness, allowing for the effects of relatives on the fitness of a target individual (i.e. neighbour-modulated fitness)? Others have concerned the extent to which the assumptions of a particular model hold for real populations. Arguments about group selection, in so far as they are not merely semantic, are mainly of this type.

Finally, are there properties that could not in principle be explained reductively? I do not think there are. I have never been very clear what an emergent property would look like. In biology, the obvious candidate is adaptedness. First, it is a defining property of living organisms (although it is one they share with their artefacts). Second, it is a property that seems to confuse most physical scientists who venture into biology without first taking the trouble to understand natural selection. Yet adaptedness, so far from being unreducible, is something that can be predicted to evolve in any population of entities with the properties of multiplication, variation and heredity.

References

Dawkins R 1976 The selfish gene. Oxford University Press, Oxford

Doolittle WF, Sapienza C 1980 Selfish genes, the phenotype paradigm and genome evolution. Nature 284:604–607

Engels WR 1989 P elements in *Drosophila*. In: Berg D, Howe M (eds) Mobile DNA. ASM Publications, Washington DC, p 437–484

Hamilton WD 1963 The evolution of altruistic behaviour. Am Nat 97:354–356

Hamilton WD 1964a The genetical evolution of social behaviour. I. J Theor Biol 7:1–16

Hamilton WD 1964b The genetical evolution of social behaviour. II. J Theor Biol 7:17–32

Hamilton WD 1967 Extraordinary sex ratios. Science 156:477–488

Maynard Smith J 1964 Group selection and kin selection. Nature 200:1145–1147

Maynard Smith J 1978 The evolution of sex. Cambridge University Press, Cambridge

Maynard Smith J 1987 How to model evolution. In: Dupre J (ed) The latest on the best. MIT Press, Cambridge, MA, p 119–131

Maynard Smith J 1988 Evolutionary progress and the levels of selection. In: Nitecki MH (ed) Evolutionary progress. University of Chicago Press, Chicago, p 219–230

Maynard Smith J, Price GR 1973 The logic of animal conflict. Nature 246:15–18
Maynard Smith J, Szathmáry R 1995 The major transitions in evolution. WH Freeman, Oxford
Nee S, Maynard Smith J 1990 The evolutionary biology of molecular parasites. Parasitology 100:S5–S18
Orgel LE, Crick FH 1980 Selfish DNA: the ultimate parasite. Nature 284:604–607
Sober E 1987 Comments on Maynard Smith's "How to model evolution". In: Dupre J (ed) The latest on the best. MIT Press, Cambridge, MA, p 133–145
Sober E, Lewontin RC 1982 Artifact, cause, and genic selection. Philos Sci 49:157–180
Williams GC 1966 Adaptation and natural selection. Princeton University Press, Princeton, NJ
Wilson DS 1975 A theory of group selection. Proc Natl Acad Sci USA 72:143–146
Wynne-Edwards VC 1962 Animal dispersion in relation to social behaviour. Oliver & Boyd, Edinburgh

DISCUSSION

Gray: I recall that Wynne-Edwards' book was brilliantly written and was full of marvellous examples. If you look back at those examples now, how many of them can be given a good explanation, and when you give an explanation what level does it usually occupy?

Maynard Smith: I think all of them can be given an explanation, and that explanation will always be at the individual organism level and not at the species level. Indeed, the reason why Wynne-Edwards' ideas were abandoned had nothing to do with the model I described. Rather, it was because David Lack and other naturalists pointed out that the traits that Wynne-Edwards described could be explained satisfactorily by individual selection.

Gray: Are you excluding kin selection?

Maynard Smith: I am including kin selection as a sort of branch of selection at the level of the individual.

Gray: Would you need kin selection to account for the great bulk of what Wynne-Edwards described?

Maynard Smith: No, I don't think so. Wynne-Edwards in the main was interested in things such as territorial behaviour, which had influences on population regulation.

Gray: The part of Wynne-Edwards' story which I found most fascinating was the bit about intrinsic population control, because there are good laboratory demonstrations that populations do stabilize long before there is a shortage of food or space. How would you now account for the population density controls that are intrinsic to the group?

Maynard Smith: Most of the stuff that Wynne-Edwards has talked about was in relation to things such as territorial behaviour. There is a good individual explanation for territorial behaviour: I will try to defend a territory big enough to feed myself, and I will exclude you from it. In doing this I'm not thinking forward to what will happen to the population later on. Basically, it can all be explained at an individual level, and that would certainly be the view of most ornithologists and naturalists today.

Bateson: I remember you saying that group selection could work if cooperating groups died out more rapidly than individuals and if there wasn't much emigration from one population to another. I wondered whether you thought there were cases where that might be true, such as in human evolution where small populations of humans were wiping each other out and there wasn't very much movement between populations.

Maynard Smith: The trouble about human evolution is that we don't actually know what the human breeding system was. Did our ancestors live in groups? How often did they move between groups? In most other primates there is actually a lot of movement between groups. Usually, members of one sex simply leave the group they were born into and join another one, so there's a hell of a lot of gene mixing.

Bateson: That may not be as true as was once thought. There may be meta populations and they move from one to the other backwards and forwards.

Maynard Smith: Later in human evolution, of course, cultural memes are going to be crucial.

Bateson: What do you think about symbiosis, where there are emergent properties of combinations of organisms?

Maynard Smith: Symbiosis is a fascinating subject. Darwinists have tended to ignore it until relatively recently because it's a little puzzling.

Bateson: Could you imagine there that there might be properties of conglomerates of organisms on which natural selection can act?

Maynard Smith: The hard question on symbiosis concerns whether we are looking at cooperation or slavery. Very often it involves slavery: for instance, the fungus has captured the alga in the lichen association.

Williams: With J. R. R. Frausto da Silva, I have written a book called *The natural selection of chemical elements*, which describes the influence of the changing environment on the elements (Williams & Frausto da Silva 1996). As we trace the history of organisms, we find the elements used has changed. An interesting example, which applies to symbiosis, is that virtually no nitrogen is fixed except by molybdenum in all living systems. This reaction system is not used except in prokaryotes, which are primitive, and the plant world apparently became dependent on the prokaryote because it cannot or does not use molybdenum in this way. As the environment changed and a whole new range of elements became available, all sorts of elements changed their functions. This made possible the advance from prokaryote to eukaryote and from unicellular to multicellular organisms, leaving each dependent on the previous system for essential functions. In this way, some spatial complexity (variables) increased while decreasing chemical complexity. The potential for this change was created by the changing possibilities of chemistry on the surface of the earth. The increase in spatial complexity probably necessitated symbiosis, simply because as the number of variables in a system increased (eukaryotes) it is of greater survival strength to leave to the more primitive organisms (prokaryotes) the simplest tasks which everybody shares. So in the end evolution has produced humans who require a whole host of chemicals from outside in order to exist. We can't even make amino acids and vitamins. The whole of evolution now becomes a very strange interdependent activity of different organisms with extreme complexity. We can see that by finding an organism to fix nitrogen, using molybdenum and keeping that organism, evolution produced man in an ecosystem dependent on a huge number of variables. I would even describe an individual human as being an ecosystem. The individual man as a human genome identity is not reproducible since he belongs to a complex history of dependent organisms and a series of complex combinations with other organisms in order to survive.

Rose: I'm cheering John all the way along on his levels of selection. I want just to make two minorly critical points. The first is, in your comments on Gaia you said that the earth cannot evolve because evolution means natural selection by means of differential reproductive survival. Technically, the word 'evolution' means change over time, and to insist that there is only one mechanism is overly restrictive.

Maynard Smith: This is just a semantic point about how we use the word evolution. I can't impose rules, but I only use 'evolution' to mean biological evolution, by natural selection or by drift, for example.

Rose: With regard to your comments on the selfish gene, you say 'of course we all know that genes aren't selfish because they are not conscious'. However, you are underestimating the organizing power of language on the way that we think about and discuss topics. I don't only mean the way that this happens in vulgarization in popular scientific writing. I think it even affects the way we think in our own laboratories about the way that the world is organized. We must be very careful about our language. You and Richard Dawkins are both very precise about refuting the 'gene for something' approach, yet that is the way one speaks about them, and the fact that this is only shorthand is often forgotten and submerged in the way in which both the language and the experiments are designed. I think this is a problem that we must be conscious of.

Maynard Smith: I think the discipline is provided by mathematics. I am prepared to think as loosely as necessary to give me an idea when I'm confronted with a new biological problem. If it helps me to think of why a gene in a mitochondrion is doing something to say, 'If I was a gene, I would do so-and-so', then I think that is OK. But when I've got an idea, I want to be able to write down the equations and show that the idea works. Of course, the equations must specify precisely what it is I'm assuming of the entities. So I think that mathematics is the discipline that one has to impose on loose thinking, but I'm all for loose thinking: we all need ideas.

Wolpert: When you started saying, 'look ultimately the way I'm choosing what I'm doing is what will give me reasonable mathematics', the issue of what you look at is becoming a question of taste. It is not about limits but it's about 'manipulability'.

Rose: There is a very important issue here, and it came up in a debate between Steven Hawking and Roger Penrose, in which Hawking accused Penrose of being a platonist: that is, these models are real because they fit the real world. Hawking says a mathematical model is a model — it performs, and that's all we need for it. In conversation the other day John was saying that this latter view is how he sees mathematical models: they are simply ways of understanding the world, but not as it were platonically, meaning that this is the way the world is.

Gray: John, I thought you were saying something more than that. I thought you were saying that natural section does work by virtue of conflict between variable organisms and that conflict, as a real phenomenon, operates at certain levels. For example, the intragenomic conflict you mentioned is clearly different from the conflict between individuals and possibly the conflict between groups (if there were to be kin selection operating at a group level). I thought you were saying something much more than that you can just model it at different levels: that there is indeed real selection at different levels.

Maynard Smith: That is a question about what the world is like. The question of whether there is group selection or not is a question of the nature of the world. Whether or not you should use inclusive fitness or gene-centred models is not a question of what the world is like, it's a question about what is a convenient way of modelling it. Both these kinds of issues get confused.

Brenner: I wanted to make a comment on this 'gene for something' concept. In classical genetics you could never assert the existence of a gene for anything until you had a mutant allele. For instance, you could not say there are genes for tallness until you had found mutants that suffered from a lack of tallness. It is different now because we can look at the genes directly. This brings us back to the question of mapping. Let us suppose we knew nothing about sickle cell anaemia. We would begin with genes for malaria resistance, that is, dominant alleles of some gene which confers the phenotype of resistance to malaria. It is another job to ask how that gene works. You would have to learn a lot of things about how malaria grows in red cells and so on, and you would be finding out how this piece of DNA maps onto the phenotype. This mapping might be direct and simple. Other genes have to operate a considerable number of levers in the way they alter the phenotype. So it's not loose talking, but there is a framework in which if you want to be rigorous that's what you have to do.

Rose: Is it not the case that a mutation tells you what the system does in the absence of the gene, rather than what the gene does? This is an important distinction.

Brenner: No, you make a mutant allele and then you infer the existence of the wild-type.

Rose: But the inference suffers from what Richard Gregory once called the 'transistor fallacy'. That is, if you remove a transistor from a radio it then emits a howl. But you can't infer from this that the function of the transistor is as a howl suppressor.

Maynard Smith: Nobody is confused by this issue. It is all quite simple.

Rose: Knockout animals are extremely important in the interpretation of neuroscience. But the whole history of knockout technology shows precisely this confusion in inferring the function of the knocked out gene from the response of the organism to the absence of that gene, granted developmental plasticity and redundancy.

Wolpert: No, I don't agree.

Dover: To Richard Dawkins' credit, he himself spotted a major flaw in the idea of the selfish gene. He recognized that if you are going to put up the idea that the phenotype is just the vehicle controlled by the replicator (the gene), it is clear that we cannot divide individual phenotypes up into 70 000 separate vehicles, each responding to the replication needs of, say, 70 000 genes. This comes back to the mapping problem between genes and phenotypes. He recognized that there is such a thing as one coherent individual phenotype, not 70 000 mini-phenotypes. In order to solve this problem of his own making, wrote a paper called *The paradox of the individual*: the 'paradox' arising from his starting premise that selfish genes are the units of selection. He 'solved' this pseudo-problem by saying that each gene draws up a list of *desiderata*: what it would like the individual organism to be doing to ensure the gene's replication. Then, all the genes compare lists and arrive at one shared list which contains some but not all of each of the separate lists. So the organism is a reflection that minimum set of items which each individual gene has proposed. Each gene doesn't get its full set of *desiderata*, but only gets a subset shared by all the other genes. Q.E.D. The problem with this is that it fails to recognize that development of multicellular systems is a sort of non-linear system, and there are all sorts of ways genes affect the final phenotype: you cannot break the phenotype down into the responsibilities of individual genes because of all the interacting systems. It is the individual and its set of internal interactions which is the target of selection. We cannot turn the process inside out.

John, when you said it was easier to model sickle cell anaemia to your students on the basis of the gene and not the phenotype, is that just a ploy to teach undergraduates, or is it a real view of what's going on in the world? The danger is that in the real world out there, many new disciplines have arisen, such as evolutionary psychology and Darwinian medicine, that are naïvely influenced by Dawkins, by the view that there is a gene-centred world out there and that every 'bit' of phenotype has its individual selfish genery story.

Maynard Smith: I don't see the problem in that way. I don't think there's any disagreement about what it is that the sickle cell gene does. In the case of sickle cell anaemia, the problem is actually so trivial you can solve it either way. One is marginally simpler than the other but they're both trivial. In other cases, I agree with Bob May when he said that he is an 'anything goes' guy: you do science in any way that solves the problem. You don't start with some general overarching principle that science is done in a certain way, you do it in the way that works.

Wolpert: Steven Rose, is that what you mean by pluralism?

Rose: Yes, there are many different ways of explaining a phenomenon of interest to us in science, and we can select from those depending on the purposes of the explanation. Therefore the way that the physiologist wants to explain something and the way the biochemist wants to explain something are both equally legitimate. This is what I mean by pluralism.

Nurse: I wanted to add my bit to the debate about mapping and phenotypes. We have to be clear that a phenotype is generally described at the level of an organism, and you can describe different phenotypes resulting from the same mutation, depending upon what you wish to describe. Let's take the simple example of an auxotrophic *E. coli* mutant that can't make tryptophan. You can describe the phenotype in terms of an alteration of the protein, which is a direct one-to-one mapping. If we then move up a level, the mutation will affect the ability of the organism to make tryptophan, which is another level of phenotypic description that is more interactive because it depends on other things that are going on in the cell. At an even higher level, the growth rate of the organism will be changed, which is even more interactive. This in turn will affect other things like the ability of that organism to survive in stationary phase. It's important to be clear about the aspect of the phenotype being considered, because often we view the phenotype as the total inclusive phenotype, which is not so easy to do when all the interactions involved are considered. When we consider selection, it is not always clear which aspect of the phenotype is important. In the *E. coli* tryptophan mutant it could be the growth rate or the ability to survive stationary phase, for example.

Maynard Smith: I strongly agree: it is for this reason that I said that fitness was specific to an environment.

Kerszberg: I would like to come back to the issue of the 'manipulability' of a model raised by Lewis Wolpert, because it points to the limits of what we can actually do. Putting things in mathematical language is a very powerful way of eliminating wishful thinking, because you make reasoning systematic and your wishes do not come into the equations. Our power to manipulate models has increased tremendously with digital computers, and I think many of the problems we have discussed stem from this increase in computing power. The number of genes that are now stored in databanks, their various interactions—

the study of all this has been made possible and is becoming a problem because of digital computers. I once went to a lecture by Garett Birkhoff, the great biomathematician, who said 'there are dinosaurs like me who are working with pencil and paper, and then there are monkeys typing at the keyboard', implying that the dinosaurs were doomed but that the monkeys can still be improved upon. What do you think of this issue of computer modelling?

Maynard Smith: I'm so bad at mathematics that I have to be able to program computers. Anybody who can't do algebra is deeply dependent upon them. Nevertheless an analytic model does give you an insight that a computer simulation doesn't.

Holmes: At the beginning of your paper you mentioned that your perspective would have been quite different had you been working in the USA. Could you elaborate?

Maynard Smith: Hamilton, myself, Dawkins and so on were all thinking in terms of gene frequencies and the sort of Haldane–Fisher type models that we were used to. The people in the States who took up the problem, tended to formulate it in terms of variance–covariance matrices and things of this kind, much more analogous to classical thermodynamics. You couldn't see the genes for the variances. I have to say that I find this stuff impenetrable, in much the same way that I find classical thermodynamics impenetrable. I can use it for working out a steam engine's behaviour, but I don't understand it; it doesn't give me an insight into what I think is going to happen next. Unfortunately, they used all the words we had been using but in totally different senses. So the term 'group selection', which had a perfectly clear meaning in the debates in Britain between 1961 and 1975, was used in a completely different sense between 1975 and 1985 by young people in America, which was a damn nuisance. We had to go over all sorts of old vomit again.

Quinn: This brings up the matter of pluralism and its relation to language. Different disciplines have different internal models, and often different languages for describing a fairly self-coherent body of information. That can't be helped. There are disciplines that disappear because their language is too cumbersome for people to devise new experiments in them. Some of this is a matter of Darwin-like selection among different scientific notations for reality, with human comprehension acting as the agent for selection.

Reference

Williams RJP, Frausto da Silva JRR 1996 The natural selection of the chemical elements. Oxford University Press, Oxford

Final discussion

What are the limits of reductionism?

Wolpert: Since we are a diverse group, it would be helpful after all this discussion to know what each of you now thinks about reductionism in biology.

Henderson: The title of this meeting, 'The limits of reductionism', implies this is a very important part of science. It is, and we all do it. But having listened to the contributions in this meeting, my feeling is that more important are factors such as levels of interest, novelty, excitement, discovery and invention. I would like to finish this symposium by asking the question, 'what proportion of all the Nobel prizes could be described as reductionist descriptions of their area?' I think the answer would be less than half.

Quinn: Mark Twain once said about the weather 'everybody talks about it, but nobody does anything about it'. It seems to me that the inverse applies to reductionism: everybody does it, but nobody wants to talk about it!

Bray: In the circumscribed arena of molecules, I think the limits of reductionism are to be found in the Third Decimal Place. You can explain how a polypeptide chain folds up to form a protein, or how proteins come together to form a cellular system in principle, but in practice we just don't know enough about them to achieve the synthesis.

Holmes: A circumspect Cartesian view of cell biology is 'give me an inventory of all the components of a cell and I will predict the behaviour of that cell'. I would question the extent to which even this limited programme is an intellectual vanity.

Bateson: I think the limits of reductionism (or at least the top down reductionism we've been talking about) are determined largely by taste and partly by being able to go no further.

Williams: The limits of reductionism are given by the knowledge of the essential units and variables of the system under discussion.

Ashmore: The limits of reductionism are set by our ability to choose or even develop (in a computer science sense) the appropriate high level language.

Morgan: I would like to echo what Denis Noble said earlier: it is good that there should be tensions about reductionism, and it is good that people working at a molecular level should feel obliged to explain the application of their work to the whole organism. It is also healthy that systems people should try to convince molecular biologists. The kind of antireductionist who is a real menace has happily not been represented here: this is the sort of person who says, 'I'm working on cognitive psychology (or whatever), so I don't need to know any molecular biology'.

Kerszberg: If I feel that some tool helps me understand a problem, I don't care whether you call it reductionist or not.

Maynard Smith: I agree exactly!

Nagel: When it comes to the limits of reductionism, there are really two questions. The first is whether there are limits to the usefulness of reductionism, but that is a red herring. There is no question that different degrees of reductionism are useful for different purposes. I remain interested in the second question: whether there are limits to the possibility of reductionism in the description of reality. I think that reductionism is impossible only if there's a big conceptual gulf of some kind. For example, I think that consciousness, value and mathematics are not reducible to physics.

Burgen: I think reductionism depends entirely on the problem. We've seen a variety of problems here, some of which are reducible. Some of them, such as the problems in my own field, are reducible until the limits of physics.

Rose: The attraction of reductionism derives from the cultural and social history of science in Western society over the last 800 years. I'm concerned about reductionism as an ideology, and I am concerned about the nature of the relationship between levels of explanation, which I don't think will be resolved. Finally, I believe that we live in one world that is, as the philosopher Mary Midgely would say, a big one: this means ontological unity and epistemological diversity.

Mitchison: Having been confirmed in my belief that one can in principle build up systems with a large number of different components, I have to confess that I find that prospect rather boring. At the end of the 19th century people thought the future was more of the same, and as we approach the end of the 20th century I rather hope that will not now also be the case.

Barlow: I suppose I have swung from being highly pro-reductionist to being slightly anti-pure-reductionism. I feel that several issues have come up here that indicate that there is a limit to reductionism and we should look to higher levels of organization more than we generally do.

Dover: I think the limits of reductionism are proportional to the magnitude of our ignorance. Reductionist explanations observe; we have imagined we have explained something merely by describing its parts, but all we have done is create an excuse not to think about it—not to think deeper about the multitude of evolutionary forces (natural selection, neutral drift, genomic mechanisms of molecular drive) that give rise to the dog's breakfasts we call biological functions.

Hess: The reductionistic approach in science is the only safe route to come to a quantitative and qualitative understanding in biology. I do not see any limits of the application of this concept to biology, as long as novel mathematical tools, biological and biochemical innovation of techniques as well as new general concepts are brought about. We know that the structure and function of biological entities is inscribed in the genome, as a prescription for generating living states. The general rules for the management of this prescription in evolution and development, in the maintenance of living states, and in timing and spacing, have yet to be explored. There might be a limit to understanding the human brain, because for reason of logic we cannot resynthesize ourselves, although we might computer-model parts of brain function, and learn how information packages are shuffled around the brain for processing, storage and retrieval, but how do we learn how we think without any syntax? No doubt, the description of nature is the prerequisite of understanding.

Brenner: We may be able to formulate limits analogous to Gödel's theorem which says there is a limit of reduction of truth to formal systems. Thus is there anything that is so complex that although we would know the answer we would be unable to formulate it in a formal mathematical language? There is always some practical limit, which is the limit of computation that you can perform.

Nurse: I agree with much of what has been said. The reductionist programme for cell biology is to reduce behaviour to molecular interactions. This isn't always possible for practical reasons, and sometimes it's unsatisfying because we don't actually get much insight into the phenomena involved. I am concerned that we will all get bored if we have to go to that level all the time, and I'm rather worried that what will happen in biology will be perhaps what happened in organic chemistry after the 1920s or thereabouts, that is, a tendency to become buried in detail. My second point is that when we are interested in pathology, and particularly in engineering (i.e. trying to manipulate the system) we certainly will have to use such a reductionist route because we need specific descriptions to be able to engineer, but that doesn't necessarily mean that this approach gives us a better understanding.

Garcia-Bellido: Reductionism versus holism (or antireductionism) is one of the many dualisms that the Western world has been using for a long time. Under scrutiny these dualisms are disappearing one after the other, resolved into fewer and fewer real questions. This happened with the notion of epistemological reductionism, that we have to have a hypothesis at one level, which means a global vision at that level, and looking for explanation in terms of the elements below, those properties which are enough to justify the properties at the higher level. So they are different faces of the same coin: both are necessary, one to think questions up, the other to answer them down.

Gray: I agree with Thomas Nagel that the interesting question is the reality of what is out there, rather than just that there are many ways of doing science. I would share in what I think is probably the common view in this meeting, that whenever reductionism is possible it is good in principle. But, on the other hand, we've heard clear examples where we must work also at a systems level. Bob

Williams has mentioned thermodynamics. Another clear exception is one which was mentioned early on in this meeting: you can only allow for natural selection to occur precisely because the laws of chemistry allow degrees of freedom so that DNA bases can be put together in different ways, otherwise you couldn't have an information programme. So we know there are clear examples where you can't reduce everything down to the elements, you have got to take in system properties. I finish by asking a question: why, then, do we care so much about it? What has barely been raised at this meeting is the underlying ideology. Steven Rose has brought up his ideology, but our chairman hasn't. Lewis, tell us why you care so deeply that 'reductionism' should always win!

Wolpert: Because I think that there is no good science that doesn't have a major element of reductionism in it, and holism is dead. Reductionism has been amazingly successful. The limits of reductionism are really only a reflection of our ignorance. The world, however, is a complicated place, and if you think something cannot be understood, my advice is 'be patient', because a reductionist explanation will eventually emerge. It may, however, take a long time with systems as complex as the brain and may require quite new concepts and technologies. We may even have to change or at least confront what we mean by understanding in biology.

Index of contributors

Entries in bold indicate papers; other entries refer to discussion contributions.

Subject index